THIS BUSINESS OF EXPLORING

By ROY CHAPMAN ANDREWS

On the Trail of Ancient Man

Ends of the Earth

This Business of Exploring

ROY CHAPMAN ANDREWS IN THE GOBI DESERT

THIS BUSINESS OF EXPLORING

BY
ROY CHAPMAN ANDREWS
DIRECTOR, THE AMERICAN MUSEUM OF NATURAL HISTORY, AND
LEADER OF THE CENTRAL ASIATIC EXPEDITIONS

Fully Illustrated

Originally published in 1935

ISBN: 978-1-940001-12-8

Published by Sportsman's Vintage Press
2013

To

My Best Editor

MY WIFE

Wilhelmina C. Andrews

Parts of this book have appeared in *The Saturday Evening Post, Cosmopolitan, This Week* and *Natural History*. My thanks are due to the editors of these publications for permission to republish this material.

R. C. A.

PREFACE

I ALWAYS like to shift the responsibility for a book of mine to some one else, or at least to divide it. The chief responsibility for this volume is my wife's, for it was her suggestion that I write an informal account of my conception of modern exploration—the exploration of today and tomorrow—as I have told it to her.

There was a selfish motive, too, for it may help to answer the questions which come to me by the thousand:

"How can I be an explorer?"

"What remains to be done in exploration?"

After telling in the first chapter what this business of exploration has become, I have amplified some of the details by my own experiences in the field and out of it.

As a concrete example of modern scientific exploration I have given an account of the 1928-30 Central Asiatic Expeditions in the Gobi Desert which has not been published in popular book form. These Expeditions were conducted with a background of war, banditry and politics in China, which made them interesting but exceedingly difficult. The story of the Central Asiatic Expeditions from 1922 to 1925 has been published in "On The Trail of Ancient Man" (G. P. Putnam's Sons, 1926). This book completes the popular narrative. The scientific record is still appearing in a series of twelve quarto volumes and hundreds of separate scientific papers.

ROY CHAPMAN ANDREWS

Oscar Seagle Colony, Schroon Lake,
Adirondacks, New York, August, 1935.

SCIENTIFIC AND TECHNICAL STAFF OF THE CENTRAL ASIATIC EXPEDITIONS 1928–1930

ANDREWS, ROY CHAPMAN—*Leader and Zoölogist,* 1928-30

GRANGER, WALTER—*Chief Palæontologist, Second in Command,* 1928-30

GARBER, A. Z.—*Surgeon,* 1930

GRABAU, A. W.—*Research Associate,* 1928-30

HILL, W. P. T.—*Topographer,* 1928

HORVATH, G.—*Motor Transport,* 1928

PEREZ, J. A.—*Surgeon,* 1928

POND, ALONZO W.—*Archæologist,* 1928

SHACKELFORD, J. B.—*Photographer,* 1928

SPOCK, L. ERSKINE—*Geologist,* 1928

TEILHARD DE CHARDIN, PERE—*Geologist,* 1930

THOMSON, ALBERT—*Assistant in Palæontology,* 1928-30

WYMAN, W. G.—*Topographer,* 1930

YOUNG, J. MCKENZIE—*Chief, Motor Transport,* 1928-30

ERIKSSON, JOEL—*Agent in Mongolia,* 1928-30

CONTENTS

Contents

LIST OF ILLUSTRATIONS

FOREWORD

Why I Am An Explorer

I WAS born to be an explorer. There never was any decision to make. I couldn't do anything else and be happy. As a little boy every moment that I could steal from school was spent in the woods along the banks of Rock River in Wisconsin or on the water itself. Wanderlust was in my blood. The desire to see new places, to discover new facts—the curiosity of life always has been a resistless driving force in me.

The wild desire to go was stronger than anything else. I had to go. I couldn't help it. My job has taken me to the Gobi. It is one of the greatest and most arid deserts in the world; a land of desolation, of thirst, and bitter cold and parching heat; of sandstorms and of tragedy. There, hardships are one's daily portion. I have been so thirsty that my tongue swelled out of my mouth. I have plowed my way through a blizzard at fifty below zero, against wind that cut like a white-hot brand. I have seen my whole camp swept from the face of the desert like a dry leaf by a whirling sandstorm. I have fought with Chinese bandits. But these things are all a part of the day's work.

You wonder why I did it; if I thought I ought to so that the world might be richer for the discovery of a dinosaur egg or a Baluchitherium or a new lake or mountain! Perhaps you think that I feel like a martyr, willing to sacrifice my flesh and health on the altar of science that my fellow men may learn the secrets of nature.

Well, I wish I might claim such lofty motives; but I'd be lying if I did. The truth is I did it because I couldn't help it; because I wanted to do it more than anything else in the world; because it made me happy; because to sit behind a desk day after day and year after year would be a greater hardship.

Until last year, for twenty-five years, I had not stayed twelve consecutive months in any one country. My home was wherever I hung my hat. I remember I said that one night to Mac Young when we were hunting Mongolian big-horned sheep on the summit of the Altai Mountains in the heart of the Gobi Desert. We had left our sleeping-bags in a saddle of the ridge while we watched a herd of sheep from behind the shelter of a rock. Not a breath of wind had stirred the grass until just as the sun sank below the western peaks; then the evening breeze came lazily up the valley, played for an instant among the rocks and passed over the crest. Instantly there were startled snorts, a rush of feet, and the hillside lay empty in the twilight shadows. The abrupt ending to the peaceful scene left us thoughtful. We lighted our pipes and wandered slowly back to the grazing ponies.

"Let's go home," I said to Mac, and, as we rode on in the stillness of the summer night, I thought of what the word had meant to me since I had begun to wander. Tonight "home" was the spot where we had left our sleeping-bags on the saddle between the peaks! In the painted desert of Gobi; in steaming Borneo jungles; among palm trees on the enchanted islands of the East

Indies; in the wilderness of Korean forests; on the summit of the Himalayas; along the fog-bound shores of Bering Sea—wherever I made my camp fire, there was "home."

Once I almost stayed a full twelve months in New York, but a robin spoiled it all. I had a penthouse on the roof of a building overlooking Central Park. My apartment was filled with Chinese things, paintings four hundred years old, a beautiful gold lacquer chest made for a Ming Emperor; silks and embroideries from the Forbidden City of old Peking. The roof was a lovely garden, bright with flowers and shrubs and even a fountain. Early one morning I was asleep in my bedroom. It was spring and the door was open wide; sunshine poured in like a golden flood. Vaguely in my dreams I seemed to hear a bird's note. It was clearer as I slowly came to consciousness, a rippling flood of song. I thought I was in my tent in the desert, but my eyes opened to my New York bedroom. I got up and went quietly to the door. Just outside on the edge of the fountain sat a robin singing his heart out. And he sang my heart out too. For suddenly I realized that it was spring; that the birds were flying north; that long black lines of geese were streaking the sky over the desert in Mongolia; that the rivers from the mountains were running full. And I was in New York, held by the invisible threads of civilization. I couldn't go. Thousands of dollars in lectures; a book to write; engagements for weeks ahead.

But couldn't I go? What use to me of the money if it didn't make me happy? How could I write a book if I were miserable doing it. Were any engagements too important to be broken? I knew it was no use. I was in the grip of the old wanderlust which has ever drawn me out upon the open road. For six months it had been dormant, but only dormant. The robin's song had waked it from its long winter sleep. The next hour I was in my office at the Museum sending telegrams and cables across half the world—"Unavoidable circumstances; war in China has unexpectedly called me back to Asia! Must sail at once!"

A few weeks later I was on the edge of the Gobi. The long brown line of loaded camels was already winding its way across the desert to our rendezvous at the "Valley of the Jewels." The first night in camp I was supremely happy. I lay in my fur bag looking at the blue cloth above me, at my rifle and revolver in the accustomed place slung to the tentpole, at field glasses and duffle sack beside my head. My old friend Walter Granger slept at my side. If it had not been for that robin singing on my roof garden in New York I would still have been there chained to the conventions of civilization.

But my job has done a good deal more than just fulfill my craving to wander. It has given me also the necessary satisfaction of accomplishment. There is a thrill like nothing else on earth in discovering something new. I had that thrill when I stood on the summit of a mountain gazing over a country which no white man had ever looked upon before. I had it out there in the desert when I realized that lying before us were the first dinosaur eggs ever seen by human eyes. I had it again in the laboratory of the Museum when we opened a block of stone ninety-five million years old from the Gobi. I had that thrill when we discovered the tomb where twenty great, shovel-tusked Mastodons had been buried two million years ago.

Reconstructing the picture; that's what fascinates me. We may not get all the details correctly, but the main facts are just as clear as though they had been written in stone and left for us to read.

It is that sort of thing which makes me know that I wouldn't trade my job for any other in the world.

But don't think for a moment that because I am a wanderer to the far places of the earth that I hate civilization. On the contrary, I like people and I like cities—by way of contrast. New York is thrilling to me: the rush and roar; the ceaseless energy; the relentless struggle for existence, the interesting people. All that is fascinating for a time. Then suddenly I must get away.

And as I stand in my roof garden and look up at the starlit sky, I see the drifting clouds and go with them in imagination far out to sea into strange new worlds. Then I count the days that still remain before I can set my feet upon the unknown trails that lead eastward to the Orient.

THIS BUSINESS OF EXPLORING

CHAPTER I

Exploritis

EXPLORITIS is not a new disease. In fact, it is just about as old as the human race. Adam's most primitive grandfather was an explorer. That's the reason why the earth today is peopled to the uttermost corners. But in most individuals the germs lie dormant, giving only a stir now and then to culminate in spring fever. Recently something has roused them to new activity, and America's young men—and women—are badly infected. I happen to know, because I am one of the physicians who are called upon to diagnose and prescribe. I suppose that "something" is the great expeditions of recent years. Amundsen, Ellsworth and Byrd in the North; the tragedy of the *Italia;* Byrd and Mawson in the Antarctic; Wilkins' submarine voyage to the polar basin; Sven Hedin in Central Asia, and perhaps, my own work in the Gobi. These expeditions all have gone into the field in the last fifteen years. They were replete with high adventure, romance and scientific achievement. Radio has kept them in touch with the world, and newspapers have told their stories day by day. Boys have learned to know the members of the expeditions as though they were personal friends. Small wonder that the urge is born to go and do likewise.

Ten years ago I used to receive an average of five

letters a week from would-be explorers. Now there are three times that number. Their tenor is all the same: "How can I be an explorer? What should I study in college? What still remains to be explored?" Fifteen a week is only my personal share. Probably the others have more than that. We might join forces and open a correspondence school. It would be useful, too, for there never was a time when exploration was more attractive than it is today. But it is not the same job that it used to be twenty years ago. The methods are different and the work itself is different. The old days of the Arctic explorer, foot slogging behind his sledge, are for the most part gone. Exploration has entered a new phase. The great pioneer lines of discovery have been thrown across the continents in every direction. Today there remain but a few small areas of the world's map unmarked by explorers' trails or where an aëroplane has not droned above the mountain peaks; only a few small areas whose topographic features are unknown.

But that does not mean that there are no new worlds to conquer. It means only that the explorer must change his methods. There still are vast regions which potentially are *terra incognita.* Many of them are mapped poorly if at all, and some hold undreamed-of treasures in the realm of science. To study these areas; to reveal the history of their making to the world of today; to learn what they can give for education, culture and for human welfare—that is the exploration of the present and the future.

Intensive exploration is just as romantic, just as alluring and almost as adventurous as that of the old days;

also it is more comfortable. Modern transportation has seen to that. Aëroplanes and automobiles! Of course you cannot stay up in the air if you are really going to find out about a country. You must get your feet on the earth and your hands on the rocks. But a plane or a motor car puts you where you want to go without loss of time or energy.

Just before I left New York for China the last time, The American Museum of Natural History was planning an expedition to the jungles of Venezuela. It is fifteen days' travel with native carriers from base to the place where they want to begin work. There is no food on the way, and the coolies must pack enough to take them there and bring them back. That reduces the load they can deposit to only a few pounds per man. Aëroplanes solve this difficulty. Two planes can load with freight, fly to the objective in three hours and land on a river. Three hours against fifteen days! Once they are in the center of the region to be explored, they will settle down for an intensive study of its zoölogy, geology, anthropology and archæology. This section of Venezuela is unmapped. The tropic jungles make travel a nightmare. But the main features can all be mapped by plane.

Their scientific exploration will tell us what is in this unknown corner of the world. A study of the rocks may yield riches in oil or minerals. New species of mammals and birds and distribution records will fill in gaps in the zoölogy of the country. They will tell us what native tribes are there, how they live and what their past has been. These are all problems which await intensive study.

The explorer of today must first of all be a specialist. Thousands of men have applied for places on my own expeditions, saying that they are "good outdoor men" but have no special training. I cannot even consider them. They would be expensive luxuries. Every white man who goes with me must do either a technical or a scientific job. All the ordinary work can be handled better and cheaper by natives. They are at home, they know the language and the customs of the country, can eat its food and endure its climate. They can be trained for any work which requires only physical qualities or ordinary intelligence.

"How can I be an explorer?"

There is only one answer: Train yourself for a technical or scientific job which fits into exploration.

Aëroplane pilots, motor experts, topographers, photographers—those are some of the technical positions that will be open on almost any expedition. Geology, meteorology, palæontology, zoölogy, anthropology, archæology and botany just about fill the bill for the usual scientific work. If you want to be an explorer go to school and stay there until you have some special knowledge that will make you useful. Just brawn and endurance won't get you very far in the exploration of today and tomorrow.

I do not mean to imply that physical fitness is not an absolute essential. I never would think of taking a man into the field who was not well or did not have ordinary endurance. I always have a surgeon on a big expedition, but for us he is there as an insurance policy. The Gobi, fortunately, is a healthy place. Gunshot wounds, broken bones and similar accidents are the greatest danger. In the tropics it is a different matter. Sickness is highly probable, fever almost certain.

From the physical standpoint it is most important to learn early what climate you are best fitted by nature to endure. There are two distinct types—hot-weather men and cold-weather men. Will Beebe is a good example of the former. He has often told me that he simply shrivels up in the cold. But he is like a salamander in the tropics. The hotter it gets the better he feels. The blazing equatorial sun brings out all his life and energy. I know a dozen others like him.

Personally I am exactly the opposite. I can stand dry heat well enough, but the wet, hot days of the tropics simply kill me. My internal motor doesn't function. I have to force myself to make the slightest effort. But in a temperate or cold climate it is a different matter. I have more energy than I know how to use, and the colder it gets the better I feel. Dryness, either in heat or cold, seems to be essential to my physical health.

Just after I was out of college, when I first began to wander, I spent a year in the East Indies. I had designs upon New Guinea, which was—and is—one of the least-known regions of the entire world. But that year proved even to my youthful judgment that the tropics were not for me. No matter how much I wanted to learn the secrets of the New Guinea jungles it could not be done. So I went to the north and stayed there, with the result that I have known hardly a day of sickness.

It is not babyish to care for your health. It is simply foolish not to do so. No explorer or any other man can do his work properly if he is half sick. Before I went into the desert I spent a great deal of time learning from a dietitian just what food was necessary to keep the men fit. I told him exactly what we could get in the Gobi, what were the conditions of climate and temperature and how we should have to live. We discussed the most minute details of our physical existence. Then he told me what a normal man must eat under those conditions. I followed his advice, and as a result we had virtually no sickness among either our foreign or our native staff. Every man on the expedition returned feeling better than ever before. Such preparedness was only common sense. We had to work at top speed from daylight to dark, and the men could not have produced the results I expected of them unless they were well fed and feeling fit.

Physical health is a prime requisite of mental health, and the latter is the most important of all. A man cannot get on with his fellows if he is mentally ill. Whenever a group of men must live together in more or less isolation for a considerable period, it is a real mental strain. It is a test of character. Small things which mean nothing in a normal existence assume gigantic proportions after a time. A man must have the mental balance to see himself in the proper perspective relative to his fellows, or else disaster ensues. Selfishness is, I suppose, one of the worst characteristics on an expedition. I would not take a selfish man, no matter what qualifications he might possess. I know several expeditions which have been wrecked because of individual selfishness. It is a thing you can't combat and is absolutely certain to make trouble.

For nerves and boredom which at some time inevitably accompany all prolonged expeditions, work is the best medicine. As long as a man has something to do he is likely to be happy. Peary told me that he used to manufacture all sorts of odd jobs to keep his men busy during the long Arctic night while he was waiting to make his attack on the North Pole in the first light of returning spring. Sir Ernest Shackleton gave me many amusing incidents of how he used to handle boredom in the Antarctic. Every leader of an expedition in almost any part of the world has the same experience under different conditions. The disease always lurks in the background, but work can usually cure it.

Two or three of my best friends, wealthy sportsmen, have begged me to take them into the Gobi on one of our expeditions. They have offered to contribute many thousands of dollars if they could only go. I have had to refuse in every case simply on the ground that there was no job for them to do. Shooting gazelles for camp meat is not very exciting, and with no definite work they would be bored to distraction. One unhappy or dissatisfied man on an expedition is like the bad apple which infects all the rest in the barrel. And if things went wrong and a little hardship had to be endured there is no compensating interest to make one forget discomfort.

It is a popular misconception that hardships are a necessary part of an explorer's existence; that without hardship and suffering there can be no real exploration! Personally I do not believe in hardships. Live as comfortably as you can while you can, is a pretty good motto. If it is possible to avoid hardships entirely, that's fine. You can do your job just that much better. But if you are in a little-known region it is improbable that you can escape without some hardship sooner or later. When it does come, if you have been fairly comfortable and well the rest of the time, you can take it in your stride and laugh while it is going on.

What is hardship? It depends entirely upon the point of view. I happen to enjoy the wild places, to love the solitude of the desert, to be happiest when I am living a primitive existence. Sleeping on the ground and eating the simplest food isn't hardship for me. Still I know hundreds of men who would simply hate it. Mac Young, our chief of motor transport, froze several of his fingers in a Mongolian blizzard. It was hardship, all right, but he wanted to go back again before the last one had been amputated. Yes, it is all in the point of view! If you want to do a thing badly enough there will be few hardships in it. If you don't want to do it everything is a hardship.

I mentioned at the beginning of this chapter that letters asking me how to become an explorer were by no means only from the male sex. Hundreds come from women, and they are the most difficult to answer. Personally I do not see just where women fit into exploration. That isn't because I am a woman hater. Far from it. I do not suppose that any man appreciates the feminine touch in most things more than I do. But on an expedition with a lot of men a woman is likely to be a liability. There are few women who are able to do technical or scientific work better than a man. No matter what one may think about equal rights, undeniably women are not as strong as men, physically. That is a serious handicap when the probable hardship comes along. Again, women are unfitted to stand certain kinds of nerve strain which is more difficult to endure than physical discomfort. They are marvelous in a crisis, but the petty anoyances of everyday life in an isolated community send them off the deep end.

One season in the Gobi we had six weeks of continual sandstorms. There never was a real calm. It simply was a matter of how hard the wind blew day after day and week after week. Most of the time it was impossible to work. We could not even read. There was sand in our food and sand in our beds. It was useless to bathe. Some of us found shelter under a bank or behind a pile of rocks. Others stayed in the tents, their faces covered with cloth. Almost every night tents were down. The continual flap and whip of canvas were maddening. After a fortnight of this every one's nerves were at the breaking point. The slightest thing might send one off into unreasoning anger. All of our staff were accustomed to sandstorms and realized that they simply had to keep themselves in hand. But it would have been ten times more difficult for a woman.

Then there is the sex problem. I do not have to enumerate the objections there. Since it cannot be solved even in our cities, why take it with you into the field? You will have difficulties enough without adding any that can be avoided. If a man wants to go exploring alone with his wife, that is quite a different matter; and it has worked with one additional male. Also several women have carried on their own expeditions successfully. But two or three women with a group of men would simply invite trouble. Others may do as they please but I'm not having any of it, thank you.

In exploration, as in most other professions, there is no royal road to success. Hard work in preparation and experience only can bring permanent results. The general principles of exploration are the same whether they are applied in the Arctic, the tropic jungles or the desert. Those principles can be learned only by experience in the field. The young man who thinks that he can organize and lead an expedition without adequate training in the hard school of experience is doomed to failure. Assuming that he has acquired the necessary technical or scientific training, his next move is to join the staff of an expedition under a competent leader. It may not be easy, but if he persists eventually he will succeed. If he happens to be connected with some museum, scientific institution, or university, his chances are that much better. His first expedition will give him an opportunity to discover many things about himself. After the glamour of anticipation and departure, when he is actually in the field and sees things as they are, he will find out if he really likes exploration as a serious profession or just as an adventure; whether he is a hot or a cold weather man, and what phase of work most appeals to him. He ought seriously to consider that home life will be denied him, and that he will always be faced with the tragedy of long separations from his family. But of course he won't think of these things if he is one of those rare individuals, a born explorer. If he is only infected with a mild case of exploritis, brought on by the universal youthful desire for romance and adventure, his first expedition may work a cure. If it does not, the road is open.

In organizing any expedition the most essential thing is a clear-cut plan of what you want to do. What is the problem? What will be its value to science, to education or to human welfare? The public is no longer interested in exploration stunts. There must be a real value to the work or it will not be taken seriously.

If an explorer has the backing of a well-known museum, geographical society or scientific institution for his projected expedition, it goes a long way in the mind of the public. They have to take it seriously even if they do not know much about the man himself. Such backing has enabled many young explorers to win their spurs. I happen to be a case in point. Without The American Museum of Natural History as my sponsor from my earliest work to the present day, I should have had a hard road to travel.

It is not difficult to interest a great museum in a really worth while project. Conditions may be such that they cannot engage in it themselves, but one is likely to obtain some support. Exploration is the lifeblood of a museum. Through it new facts and new collections flow into the institution. Exploration is what makes it a vital force in the educational life of a city or a country. But most museums are poor. I never yet have seen one which had all the money it needed for its projected plans. The American Museum of Natural History is an excellent example. Even with an annual budget of a million and a half dollars we do not have sufficient funds with which to conduct the work of the institution. Therefore do not expect that a museum will jump at the chance to finance your expedition, no matter how worth while it may be.

Finances are the *bête noire* of every explorer. How to raise the money? Unless you have the personality and ability to sell yourself as well as your plan you are just out of luck. Enthusiasm is a *sine qua non.* You can't make some one else believe in your project unless you believe in it passionately yourself. I never shall forget one day when I had been talking to Mr. J. P. Morgan regarding the Central Asiatic Expedition. I wanted to do it so intensely that I got all steamed up.

Just as I was going out Mr. Morgan put his hand on my shoulder and said, "Roy, don't ever lose your enthusiasm. It is the most valuable characteristic you possess."

Of course, financing an expedition has many different aspects. As a rule, one or two individuals give you a nest egg to build upon. It is the first donation that is most difficult to get. When you go to see some possible contributor he is almost certain to ask, "How much do you need and how much have you got?" If you can say that you already have a substantial amount it has a distinctly favorable effect. He realizes that some one else thinks your plan worth while, and it confirms his own judgment. I never obtained anything except small amounts by writing letters. It is the personal interview that counts. You must be able to infect the other fellow with your own enthusiasm, and it is pretty difficult to do that on paper.

Money raising is a hateful job. I never found any explorer who liked it. Still, if you want to run your own expedition it is one of the things that you have to face.

No one else can do it for you. A business manager can help a lot, but the brunt of it must fall upon your own shoulders. It is easier to raise money for a large expedition than for a small one. If your project is big enough it stands a better chance of appealing to men with great fortunes, for they are accustomed to dealing in big things. The small ones do not interest them and they will say yes or no without much thought either way.

Publicity is a valuable asset, of course. Also it may be very dangerous. The line between beneficial and harmful publicity is exceedingly thin. The difficulty is that publicity cannot be controlled. A newspaper reporter must have some peg upon which to hang the serious account of what you intend to do or have done. He wants something with a popular appeal, and often what you really wish the public to know is distorted or lost entirely. Over-publicity is a danger. People become bored if too much is written about an expedition. That did not happen before the days of radio. Real results are always interesting, but continual accounts of trivial happenings often defeat the purpose for which they were written.

The ability to lecture or to write entertainingly is an enormous asset. Then you do not have to depend upon a second person to state your thesis, and the public learns to know you at first hand. That, I should say, is the most valuable form of publicity, for it is one which you can control absolutely. But if you write or lecture do not under-estimate your audience. By that I mean don't think that it wants merely to be amused. People go to lectures to be instructed, but the instruction must be put in the form of a sugar-coated pill. If you can give them enough amusement to make the instruction easy to swallow, the result is perfect. They want something that they can think over and talk about at breakfast next morning. If amusement only were desired they would go to a movie or a play. I have lectured for twenty-three years and at first I made the very mistake that I have warned against. I selected the amusing or exciting episodes of my work and touched only casually upon the serious aspects. I found that it was not successful and my lecture manager told me why.

Adventures, of course, are always associated with exploration. Yet they are the one thing which a real explorer tries to guard against. My favorite quotation is Stefansson's dictum: "Adventures are a mark of incompetence." It says so much in a very few words. It means that if you have an adventurous expedition you did not prepare yourself adequately. Adventures are a nuisance. They interfere with work. There are many so-called explorers who are really travelers seeking adventure. They welcome every opportunity for a hairbreadth escape or some thrilling experience because it is their stock in trade. Then they write a book about their experiences. Not having a serious objective which gives them something worth while to contribute they tell the story of their hardships. I could mention a dozen such men. If the explorer has a clear-cut problem to solve and an honest desire to do something really worth while he will prepare against adventures.

Last winter I heard a man remark that Admiral Byrd's Antarctic Expedition was too easy to be interesting. He went on to say that it was a *de luxe* expedition and really wasn't a very difficult job. Although such a statement only exposed his ignorance it made me rise in wrath. Admiral Byrd's expedition was a colossal undertaking. It was magnificently organized and directed and was completely successful. Any one who calls an expedition of that sort easy deserves the title of fool. The fact that he did not have a series of disasters shows better than anything else what an able explorer Byrd is. As far as a *de luxe* expedition is concerned, Byrd would have been stupid if he had not made use of every modern invention in transport as well as for the comfort and health of his men during their long isolation in the Antarctic. I will venture to say that he did not take a single item of unnecessary equipment. Success in every branch of his work shows how right he was. Byrd is a modern explorer. His first care was to eliminate by thorough preparation every adventure that foresight could predict.

Because I advocate preparing against adventures I do not mean to imply that a leader should never take risks. Better not go out at all if that is his policy. It should be a matter of judgment. If the object to be gained warrants the risk involved, go ahead. Otherwise don't do it. A leader has to decide that for himself. Peary told me many years ago that the leader of an expedition must be prepared to back his own judgment against that of all his men if need be. It is his responsibility. Failure reverts on his head alone. He is supposed to have the experience and judgment to decide better than any one else. If he does not have that he isn't a good leader. After all, the principles of exploration are much the same as those involved in directing an army or in conducting a big business venture.

What remains to be done in exploration? I have already answered that question, I think. The areas of the earth's surface open to intensive exploration are legion. They exist on every continent. The Arctic and the Antarctic, Africa, Asia, South America, Australia and New Guinea are most clearly indicated. Central Asia, I believe, offers the greatest field. But it is also the most difficult to enter. Recent political events in Russia and China have erected well-nigh impassable barriers to effective scientific exploration in Central Asia.

Russia is active in exploration, but its dominions are so vast and so little known that without assistance they can make only a small impression upon what remains to be done. Although they protest that they are not adverse to coöperation with foreign scientific institutions, internal conditions make it not easy to conduct a large expedition in the regions which they control. In South America, Africa, Australia and New Guinea governmental permission to carry on explorations is not difficult to obtain. There is, of course, much archæological research to be done in many parts of Europe and North America. Enough intensive exploration remains for a hundred years of work; probably even then it will not be finished.

CHAPTER II

Danger

WHENEVER I start on an expedition the insurance company always cancels my accident policy. At first it used to make me mad; now I just laugh at the ignorance of the company and cheerfully let it go. It saves me money and they lose because I am not half so likely to have an accident on a trip of real honest-to-goodness exploration as I would be in the city. Honestly, if I had had as many narrow escapes in the Gobi Desert as I have had from being killed by automobiles or in other ways in America, I could write a whole book about them. The trouble is that such escapes become commonplace. Every one of you who lives in the big cities has them every day or two and it has robbed them of interest. I am not saying this just to pose as a man of intrepid courage, but because it is true, as any real explorer will tell you.

In the field, if your expedition is to be successful, you prepare against adventures; yet now and then one will happen in spite of foresight. I suppose you can call them acts of God. But they are exceptions and as a rule on a well-conducted expedition life goes on pretty safely. You are living much more normally than you could possibly live in the city and you are not dependent upon the devilish mechanical devices which necessarily complicate a communal existence. Just by way of contrast I'll tell you of a few "adventures" I've had in the field and at home.

One that happened in Mongolia wasn't due to carelessness or stupidity. As a matter of fact why it happened is a mystery to me today and always will be. I was driving across Mongolia from Kalgan to Urga with a friend who later was murdered in China. We were in an open Dodge touring car, traveling hard, trying to make the crossing in less than four days. Three hours at the wheel and three hours' rest was our schedule. I was driving when we reached Ude, a collection of five or six Mongol *yurts,* in the center of the Gobi. The trail ran about four hundred yards from a rocky promontory which juts out into the desert. Charlie was asleep in the rear seat and I was half dozing over the wheel for the road was good. Suddenly five men appeared on the end of the promontory and, without the slightest warning, opened fire on our car. They couldn't have been Chinese because they were doing awfully good shooting and a Chinaman is the world's worst shot. The bullets were zinging above our heads and plumping into the motor every minute. Boy, did I wake up! I humped myself over the wheel, trying to make myself as small as possible, and stepped on the gas. The speedometer showed fifty miles an hour but the bullets were still hitting us. Charlie got our guns out from under a robe and I yelled,

"We'd better do something about this or we'll be killed. Pass me my rifle."

I leaned backward to take the gun and at that very second a bullet smacked against the steering wheel, shattering the whole lower side where my body had been pressed only a second before. Call it Fate, Divine Protection or what you will—the fact remains that if I had waited a fraction of a second before leaning backward, I would have been a dead man. That bullet simply wasn't marked with my name.

The trail led round a high wall of rocks into the soft sandy bed of a dry stream. It was a bad place and we knew that we never could get through without some strenuous pushing on the car. Sure enough we were stuck in a few moments but at least we were completely out of sight of the men who were shooting at us.

Leaving the motor running, Charlie and I climbed the rocks and peeped over the top. The five men were standing in plain sight about three hundred and fifty yards away, evidently consulting about their next move. They knew of course that we would be stuck in the sand. Moreover, since we had not fired in return I suppose they thought that we had no rifles. Anyway they started to climb slowly down the rocks to come across the open plain to our car. Something had to be done about it as apparently they were bent on murder.

Charlie selected one fellow who was standing silhouetted against the sky and I lined my sights on another just in front of him. I was shooting a Savage 250-3000 with soft nose bullets and Charlie had a Ross .280. As our rifles crashed both men crumpled. The other three disappeared like shadows behind the promontory. We waited for a time but none of the three showed themselves. Then we scrambled down to the car trying desperately to push it out of the sand. It took more than an hour of hard work to get through the canyon, for every few minutes we had to climb the rocks to be sure that we were not being stalked by the remaining bandits. I don't mind saying that we had the jitters.

Who those fellows were and why they attacked us I could never find out. They were dressed in Mongol clothes but that proved nothing. That they were not Chinese we felt certain because they shot too well. They might have been Russians but I hardly think so. Probably Mongols with Russian rifles.

So much for Asiatic brigands. Let me tell you what I have seen of the New York variety. I had returned from the Orient in October feeling a little unhappy about the prospect of an unexciting winter in America.

Three days after my arrival, at five o'clock in the afternoon, my secretary at that time, Miss A. L. Seeling, left The American Museum of Natural History and walked west on Seventy-eighth Street toward Amsterdam Avenue. The street was almost deserted. Suddenly the door of a brownstone house (No. 154 to be exact) was flung open and a man staggered down the steps. He was dripping blood and carried a pistol in his left hand. A taxi dashed by. A feeble gesture brought it to the curb; the wounded man muttered something to the driver and then made his way painfully back up the steps. A moment later he returned with another man, supporting between them the sagging body of a young Italian. A dark red stream followed across the pavement, spreading into an ugly blotch as they lifted him into the car. Miss Seeling had seen enough. Weak and sick she signaled a cruising taxi to take her home.

Fifteen minutes later I left the Museum and started west on Seventy-eighth Street. A big police truck swung around the corner of Amsterdam Avenue, roared eastward and stopped at No. 154. Twenty uniformed officers, some of them carrying sawed-off shotguns, piled out. Two more trucks with sirens shrieking followed. The street was alive with police, all armed with shotguns or pistols. Four of them hustled out a wicked-looking machine gun and set it up on the opposite side of the street, facing the door of No. 154.

Then the newspaper reporters began to arrive. I saw one who had interviewed me only the day before and caught him by the shoulder as he ran past.

"What's it all about?"

"I don't know yet," said he. "We got word that a detective had been shot in a gun fight here. I believe there are three or four of them and it's likely to be a big show."

"By Jove! I want to see it," I said. "Can you give me a reporter's card?"

"Sure, you can take mine. I don't need it. The police all know me."

I stuck the card in my hat and we went inside the ring of curious spectators up to the big cars. I didn't bother my reporter friend with any more questions. He had a job of work to do and my ticket to the show got me a ringside place. This is just what it was—a show! I could hardly believe it to be a real story and not a movie as the tragedy unfolded. The action was fast and dramatic. A second machine gun mounted on the high steps of a house above and to the right of the first, pointed its black muzzle to the flat roof of No. 154. Beside it a powerful searchlight picked out every brick and stone as it played along the cornice. A cat could not have moved on the edge without being seen by the grim officer who swung the gun back and forth along the beam of light. God help the man who tried to escape that way!

In the street below, a clanging ambulance drove a path for itself through the packed line of spectators. Four policemen with a big hand searchlight, shotguns and pistols crossed the street to the front door of the house. I tried to slip in behind them but the rearmost officer caught me by the arm.

"Hey, Buddy, what's the big idea? There's likely to be some shooting in there. You get your story from the outside."

Sheepishly I grinned and backed out. The officers disappeared. After five or ten minutes, one of them came to the door and called the ambulance. The body of a man was lifted into the stretcher. I caught a glimpse of his white, blood-stained face in the light. It was not a pleasant sight.

The four officers who had gone into the house emerged through the basement door just as Police Commissioner Mulrooney arrived with six or seven plain-clothes men. He went inside with a dozen others. Fifteen minutes later two came out leading between them a hatless man in shirt and trousers. They had found him in bed in the upper story of an adjoining house. He was still protesting his innocence.

"You haven't got anything on me," I heard him wail. "I was asleep up there, dead asleep. I don't know a thing about it, so help me God! So help me God, I don't!"

Grimly silent, the officers hustled him into the waiting police car. Then an amusing thing happened.

Half a dozen newspaper photographers let off flashlights in a blinding glare. You should have seen the crowding spectators rush for cover! They thought that the shooting had begun again. Stampeding down the street like a herd of cattle, they ducked into basements and doorways and flattened against the nearest walls. From my stand on the steps, I got a bird's-eye view of the whole picture.

By that time I knew pretty well what it was all about. It seems that an Italian detective patrolling this district had recognized a gunman, Enrico Grieco, who was wanted for the murder of a policeman three years ago. He trailed the man to No. 154 and telephoned for assistance. Two other detectives, both Italians, joined him and the three started to search the house. The landlady's loud protests at the invasion of her premises warned the killer who, with two others and a woman, was in a room on the third floor. The gunmen shot through the door, badly wounding two of the detectives. The one Miss Seeling first saw had six bullets in his arms and body. The other, James Pasagno, had five. He died the next day. The officers returned their fire, killing the murderer and wounding another. The remaining two and the woman escaped over the flat roofs of the adjoining houses.

For an hour the search went on but nothing new happened. There were fifty uniformed police, a dozen plain-clothes men, two machine guns, five red service cars, gas bombs, and Heaven only knows what other gear clustered in front of that house. It looked like a miniature army. And this almost in the sacred shadow of The American Museum of Natural History where I toil and have my being! It was well-nigh incredible. Only eighteen days before I had left China with its war and bandits. Here in the so-called safety of New York, virtually at my office door, was more war and worse bandits. I must admit that I got a great kick out of it. I had been vaguely uneasy about how I was going to stand six months in the city with no real excitement. And after only three days, I was presented on a silver platter, as it were, with this movie drama of New York life.

The trouble was that I could not get out of my mind the brief glimpse I had of that white drawn face on the stretcher, or forget those wounded detectives. They were brave men, those three. How few among the careless millions in New York knew of that sad funeral a few days later in the Italian quarter where the dead officer lay in a flower-heaped coffin.

At a quarter past six I left to dress for dinner. Already on Amsterdam Avenue, where I got a taxi, newsboys were calling "Extras," describing the gun fight. I bought a paper. There it was in half a column. My reporter friends had telephoned the story from the corner drug store while I had watched the living picture unfold from my stand beside the machine gun. Just an hour and fifteen minutes since the first shot was fired! That's New York! It takes a foreigner or a wanderer like myself, who is accustomed to the unhurried life of the Orient, to appreciate and marvel at the things they do in this fantastic city of ours.

China, of course, swarms with bandits. But they are not the Chicago or New York kind. They don't shoot you in the back or pot you from an automobile with a machine gun. In the first place, Chinese brigands are still somewhat loath to kill a foreigner. They give their victims a chance, and if they can get what they want without shooting they won't open fire. But from all I can learn of our modern city gunman, he has no conscience or ethics about killing. He would kill just as soon as not, perhaps rather, and to do it in a nasty way when you didn't have a chance.

Chinese bandits annoy us every time we go to the Gobi Desert. At Kalgan, where we leave the railroad and civilization, half a dozen big camel trails converge. Brigands always concentrate in this region to rob the caravans carrying opium, furs and other valuable goods. Bands of from ten to a thousand infest the trails. They don't like to stand up to rifle fire and if one or two of them are killed in the first volley the rest run like stags. About twenty Chinese bandits to one well-armed foreigner is proper odds, as we have found by experience. Of course, when we get well out into the desert there are no brigands at all, for there is no one to rob and they cannot live off the country. A Chinese bandit sounds like a very terrible person from the distance of New York, but the closer you get to him the less fearsome he becomes. He is invariably the world's worst shot; he can't shoot quickly and he dislikes to shoot at all. He is just about the opposite in all things to the American variety. One experience I had will demonstrate what I mean.

Our camp one year was three hundred miles from Kalgan and I had to return to China for extra supplies. One of the men drove one car and I the other. Mine was a couple of miles in advance of his when I came to a deep valley where two Russian cars had been robbed only a week earlier. We had heard of it for the bandits had taken twenty thousand dollars' worth of sable skins and killed one man. The other had been stripped absolutely naked and left to find his way to Kalgan in his birthday suit.

Just before I reached the spot I thought, "I wonder if there is any chance of my being held up there. Lightning doesn't strike twice in the same place but brigands might."

A moment later I saw the head and shoulders of a man on horseback just appearing over the summit of the hill three hundred yards away. The sun glinted on a rifle barrel. Now there are just two kinds of men who carry rifles in China—bandits and soldiers—and the two are synonymous. Anyway I had no mind to have him there whoever he was. I dropped a bullet from my .38 revolver too close for comfort even though there was no attempt to hit him. He disappeared abruptly. Just then my car swept over the edge of the valley and started down the steep slope. In the bottom two hundred yards away were four horsemen, rifles on their backs. I knew instantly that they were bandits and that I was in for it. The trail was narrow and rocky and I couldn't turn. Also I knew that a Mongol pony never would stand against the charge of a motor car. Opening the cut-out I stepped on the accelerator and the car rushed down the hill at forty miles an hour, roaring like an aëroplane. The ponies went mad with fright. At first the brigands vainly tried to get the rifles off their backs, but in a moment their chief concern was to stay in their saddles. The ponies were rearing and plunging madly and three of them raced off across the plains. The fourth seemed too frightened to run.

I was right beside him and I'll never forget the look of abject terror on the face of that Chinese bandit. My revolver was in my right hand and of course I could have killed him easily but there was no use in doing that. He had on a peaked Mongol hat and I fired at it five or six times trying to knock it off his head. Finally his pony started after the others with me behind in the car yelling and shooting. We reached the rim of the valley and I let him go then. All four of them had the fright of their lives and I had had a lot of fun. When I reported the incident to the commander of a detachment of Chinese soldiers fifty miles farther on, he was furious because I hadn't killed at least one of the bandits. I told him, however, that I was a peaceable explorer and that it was his business to kill brigands, not mine.

In 1928, I was present at the annual dinner of the Wilderness Club at the Racquet Club in Philadelphia. I was to take the midnight train to San Francisco, to sail for China. On one side of me sat Colonel Theodore Roosevelt and next to him Dr. George Gordon, Director of the University Museum. As I left the table I shook hands with Ted and Gordon.

"Take care of yourself. We don't want you to be killed over there in China," said Gordon.

"I'll be all right," I laughed. "It is you fellows in the city who ought to worry."

What a tragic prophecy! Ten minutes later Dr. Gordon was dead. He left the table just behind me, got his coat at the check room, slipped on the marble floor and fractured his skull.

Think also of disease; cities versus country. The more people the more germs. No one can deny that. In the Arctic and Antarctic and in the desert, disease germs hardly exist. During all the years of my own expeditions in the Gobi we never had a case of serious illness.

In 1928, I got a bullet in the leg. For five days all the dressings had to be done in sandstorms. Dust simply couldn't be kept out of the eighteen-inch bullet wound, yet it healed perfectly with no trace of infection. Under similar conditions in a city, I should probably have died.

On Admiral Peary's last and successful attempt to reach the North Pole only once did any of the men have a cold. That was when they opened a case of books on the ship during the long Arctic night. Evidently it had been packed by some one who had a head cold. That some one probably sneezed over it a few times and deposited the germs as an additional present for the explorers. Every mother's son of them on the *Roosevelt* got it, too.

One of the narrowest escapes from death that I ever had actually occurred in my office at The American Museum of Natural History. I was sitting at my desk when the telephone rang. While speaking into the instrument I tipped back in the swivel chair and crashed to the floor. The screws had broken and the seat of the chair separated from the frame. I landed on the side of my head and the doctor said that it was the closest call to a broken neck that he had ever seen. If one of the vertebræ had slipped the smallest fraction of an inch farther I would now be pushing clouds.

Of course bathtub accidents have become a subject for the funny papers. Everybody has one sooner or later. But often they end in tragedy. In 1923, one of my most intimate friends, Clark by name, a man whom I had lived with for two years, was killed in his own tub. He slipped, fell violently against the faucet, and ruptured some of his internal organs. We hear about the bathtub accidents because they have elements of the unusual. But being hit by an automobile is neither funny nor unusual. Therefore thousands of such deaths go unchronicled in the news columns.

When I arrived in New York one year after a long absence in the Orient, I was positively frightened. The life of our city had so changed in its attempts to evade prohibition and to give lucrative employment to its thousands of gunmen that I felt like a lamb among wolves. I was told that I must not do this or that or the other thing, else dire consequences might ensue. One day I was taken to a "speakeasy" for dinner. It was my first speakeasy and I was as thrilled as a country cousin. I felt terribly naughty and excited at breaking the law in such a delightful way. And then they told me of what had happened to a mutual friend at one of the other thirty thousand speakeasies in New York City. How his liquor was drugged and how he waked up in a room with four gorilla-faced gentlemen as companions. They suggested that he sign a check for three thousand dollars which they said was the value of certain furniture and fixtures destroyed by him while under the influence of bootleg gin. The four anthropoids were willing, nay anxious, to swear that he had ruined much more property than three thousand dollars would purchase but they considerately let him off at that. He did sign and promptly stopped payment on the check, whereupon the gorillas appeared openly at his office accompanied by a lawyer of very dubious reputation, threatening legal action and other dire consequences if the check were not paid.

This was only one of similar stories with which I was regaled. By the time dinner ended, I felt so small and frightened that I wanted to take the first ship back to the Orient where I could be comparatively safe in Mongolia. Then we left the speakeasy. Taxis stood invitingly at the door. Do you think my sophisticated friends would so much as look at one of them? Oh, no. They were especially to be avoided. You might be taken for a ride in the Park where the driver had friends waiting to relieve you of your valuables.

In the winter of 1933-34 I had an amusing experience with a burglar in New York. I suppose it might be called a real adventure. I had a penthouse on the top of the Hotel des Artistes. Wallace Morgan and Pat Enright, the famous illustrators, share an apartment just below and a ladder leads up the wall from their terrace to a tiny balcony opening out of my bedroom. I slept on a studio bed about six feet from the French windows of the balcony.

Many years in the field have made me very sensitive to any unusual noise or to the presence of any one in my room. It was five o'clock in the morning just before dawn and I was in a sound sleep. Suddenly I raised up on my right elbow wide awake. There crouched a man beside the head of the bed, his face about eighteen inches from mine. He was a regular stage burglar; a soft cap pulled far down over his eyes, and a short coat tightly buttoned.

There he sat looking at me, evidently waiting to see if I were asleep before he began to prowl. For an instant I thought that I was dreaming. Then I realized that it was a man and a wave of indignation swept over me. The idea of a man sitting in my bedroom made me absolutely see red.

"You ——— ——— ———," I yelled, and jumped at him. Apparently that was the proper way to talk to burglars, for he leaped backward and made for the balcony over which he had come.

My feet were tangled in the sheets and I landed on the floor on all fours. The burglar was over the balcony rail except for one leg when I caught his trousers with my left hand, half on the floor myself. He kicked like a mule. The first kick landed on my chest; the next got me full in the face. He tore loose and ran down the wall ladder like a monkey, and into the house.

I phoned the operator at the front entrance and he sent in a call to the West Sixty-eighth Street police station. In five minutes seven policemen were at the door. The burglar must still have been in the house as he could not possibly have run down twenty stories to the basement where he had entered, in that space of time.

To my amazement all seven of the officers piled into the elevator and out upon Wallace Morgan's roof.

"What are you doing up here?" I asked.

"Looking for your burglar."

"Hell, I just chased him away. He is in the house. Did you leave any one in the basement?"

"No."

And then they proceeded to search the building from the top down! Naturally my burglar simply ran ahead of them and out through the fire door in the basement, which was the way he had entered. It was gray dawn but the man's silhouette against the sky made me feel sure that he was an Oriental. I have lived in the Far East so long that I couldn't be mistaken about that.

The next morning I had an interesting talk with a detective from Headquarters. I remarked upon the agility with which the burglar ran down the wall ladder. A monkey couldn't have done it faster.

"These cat burglars spend hours every day in gymnasiums," said he. "They keep themselves very fit and practice running up and down ladders and jumping from one ledge to another."

"Do you suppose," said I, "that he had a gun?"

"Probably not. Blackjacks are more in their line. It was lucky that you waked up suddenly and jumped at him. It was the one thing he wasn't expecting. If you had roused slowly he'd have knocked you on the head and put you to sleep for awhile."

I really think what frightened the burglar was the volume of curses which I spouted at him. As I remember, I did some very artistic swearing. They were so effective that I have thought up a new supply for the next gentleman of the profession who comes into my bedroom uninvited.

To recapitulate: I honestly believe that the average modern explorer is in less danger, by long odds, in the desert, or the Arctic or the jungle, than is the inhabitant of almost any of America's great cities. My contention is that the city dweller is so accustomed to the continual dangers which beset him on all sides every day that they have become a part of his life. He doesn't even think about them. Whereas, the dangers which an explorer meets are so *unusual* and interesting just because they are different that they seem to be greater.

Of course if an explorer does not study his problem and prepare properly before he goes into the field, something is bound to happen. He can find plenty of danger if he is searching for it. So could you if you did not look in both directions when you cross a city street.

I am a fatalist now. I believe that when my time comes I'll go and I don't spend any time worrying about it. As for danger, one should recall what Mark Twain said:

"Beds are the most dangerous places in the world because so many people die in them."

CHAPTER III

Snakes

TO MEET the popular conception of an explorer a man must have suffered cold, heat, starvation, fever, attacks from wild animals and savage natives and must have been bitten by snakes. Snakes are essential. If you haven't had snakes—real ones—you just can't be a proper explorer. I suppose it is because for many people snakes have a horrible fascination. I've had friends visit me at the Museum who said that they loathed and detested snakes. Yet they wanted first to see the reptile exhibit. Often before a lecture I have received letters from both men and women who asked if I was to show any snake pictures. If so they couldn't come because snakes made them hysterical.

But it isn't so for every one. I used to know a woman in The American Museum of Natural History who loved snakes. Sometimes she would stroll into my office with a whacking great serpent coiled around her waist and the darned thing's head tucked affectionately under her chin. Needless to say she wasn't married. Perhaps she might have been if it hadn't been for her snake habit, for she wasn't bad looking and was undeniably brilliant. But what man would care to come home at night to find a snake roosting comfortably in the middle of the other twin bed! Doubtless that is where it would have been too, for the people with a snake complex don't seem able to understand the aversion that the other ninety-nine and ninety-nine one hundredth per cent of the human race have for their pets.

I have heard that reptiles were canned for food and I knew one man personally who professed to enjoy rattlesnake to eat—said it tasted "just like chicken." Funny how animals—like monkeys for instance—which no one else likes to eat always taste "just like chicken." Anyway I've eaten plenty of chicken—what man who has lived in China hasn't—and I've also eaten monkeys and my monkeys *didn't* taste like chicken. Far from it! I ate them because I was so hungry that I'd have eaten anything—except snake.

I remember that when the late President Theodore Roosevelt came back from his expedition to the "River of Doubt" where his party nearly starved to death, I was lunching with him at Oyster Bay. Senator Beveridge asked him how he liked monkeys to eat.

"Well," said the Colonel, "you could lock me up in the monkey cage at the zoo with no danger to the inmates." He didn't think they tasted like chicken either.

Speaking of Colonel Roosevelt, the late Carl Akeley, African explorer, told me an amusing story about a dinner at the White House. The President had been having a good deal of trouble with Congress just at that time. Congressman Mann was sitting at his left. Akeley was relating an experience in Africa when he saw fourteen lions come out of a cave. The President turned to Congressman Mann and said:

"I'd like to have those lions here in Washington."

"What would you do with them, Mr. President?"

"I'd turn them loose on Congress."

"But wouldn't you be afraid that they'd make a few mistakes?"

"Not if they stayed long enough," laughed the Colonel.

But to return to snakes—personally I dislike reptiles intensely. I don't know why, but I just do. My dislike isn't fear. It is an instinctive loathing. I have had to collect hundreds of reptiles during my explorations and I can handle them, if I have to, without going into hysterics or anything of that sort. I can inject them or take out their insides or skin them but I can think of about a million things that I would rather do. I have tried to remember when I began to dislike snakes so actively because I am interested in almost all other living things, and I am sure it dates from the time when I was about fourteen years old.

I was shooting along the banks of the Rock River in Wisconsin and spent the night in the open. After my dinner of bacon and bread I curled up to sleep with my head on a coat at the root of a great tree. During the night I felt something wriggling in my hair and sleepily put up my hand. A cold body curled about my wrist and bare arm. You could have heard me yell for at least seven miles and I leaped up shaking with fright. It was only a garter snake but if it had been a rattler I couldn't have been more scared. I didn't get over it for weeks. I used to have snake dreams even though I had never even smelled the cork of a whisky bottle at that tender age.

Snakes don't get on my nerves to that extent now but I'm still pretty jumpy when I know they are about. One year in the Gobi Desert the Central Asiatic Expedition was actually driven away from a most productive fossil field by a plague of poisonous vipers. That wasn't just a case of nerves—we had a jolly good reason for leaving. If you had killed forty-seven snakes in your tents in the course of a couple of nights, I think you'd leave too—unless you were like my "girl-friend" who doted on them. She would probably have enjoyed herself so much that she'd be there yet. That year our camp was pitched on a high promontory which jutted out into the desert like the prow of an enormous ship. On the very end was a great pile of rocks, a religious monument which the Mongols call an *obo*—built to the spirits of nature.

There was a temple about four miles away on the basin floor and a few hours after our tents were up three or four lama priests rode over to call. They politely asked if we would please not kill anything there because it was very sacred ground. Of course I promised, not thinking of snakes, and they departed happily. Fossils were abundant all along the sides of the promontory and each man had located a valuable specimen within a few hours. Also each man reported vipers when we gathered for dinner in the mess tent that night. Some had killed four or five, others only two or three, but all had seen snakes.

So far as I am aware, there is only one poisonous reptile in the Gobi. The desert is too dry and the climate too cold. This one is a pit viper, about the size and shape of our copperhead. He is pretty bad medicine too. I don't know whether or not he contains enough poison to kill a healthy man, but a good bite would make even Jack Dempsey pretty sick.

We got along well enough for a few days but one night the temperature suddenly dropped almost to the freezing point. The vipers evidently didn't like it and made tracks for our camp to get warm. What snake instinct told them that our tents were there and that in them they could find warmth and shelter I can't imagine. Be that as it may, they wriggled their fat loathsome bodies out of the rocks, up the sides of the promontory and into our camp. They came not in twos or threes but in dozens. The first we knew of it was from Norman Lovell, one of our motor engineers. It was a clear brilliant night and he waked about two in the morning just in time to see a huge serpent wriggling across the patch of moonlight in his tent door. Lovell had no mind to share his house with a viper even though it was merely trying to get warm, and reached for his flash lamp to find his shoes. Before getting up he thought that it might be well to have a look about. Sure enough, coiled on both legs of his cot bed were two other snakes. He got his small collector's pickax, gently untangled them and cut off their heads. With his shoes on, he began a still hunt for the original visitor. Just as he stepped on the ground the grandfather of all vipers crawled out from a gasoline box at the head of his bed.

Meanwhile sounds of dismay were issuing from the abode of Fred Morris, one of our geologists.

"Dear God, my tent is full of snakes," I heard him moan. "There are hundreds of them."

From Mac Young came a volley of curses. "You yellow b——d, get out of my shoe!"

Mac had waked just in time to see a viper crawling into one of his boots. The whole camp was astir and every one was having a lively time. One of the Chinese chauffeurs found a snake coiled in his cap. Another was actually in the cook's bed. Not only were there dozens of them in camp but more were coming up over the edge of the escarpment.

True to their religion the Mongols wouldn't kill one of the snakes on the promontory. They contented themselves with shooing them out of their tents and calling to one of us to do the dirty work. They weren't taking any chances with the Mongol gods as long as a heathen foreigner was on the spot to accept their responsibilities.

I had put on only tennis slippers, for my leather shoes were at the bottom of a duffle sack. With a flash lamp in one hand and a pickax in the other I stepped out of the tent door onto something round and hard. I must have jumped three feet straight up and what I yelled made blue sparks in the air. But it was only a piece of half inch rope! A moment later Walter Granger made a vicious lunge at a pipe cleaner!

All of us had the jitters. There was no more sleep that night and daylight found us making a complete overhaul of everything in our tents. Snakes were everywhere—in gun cases, in duffle bags, under blankets. Fortunately, however, they were chilled and consequently sluggish. None of us was bitten but we certainly wouldn't have escaped if we had waited until the sun had warmed the tents. Wolf, my police dog, was the single casualty, but he was struck by such a small snake that he escaped with only a few hours of sickness.

We remained there two more days but the supply of snakes was inexhaustible. After we had killed forty-seven it seemed like a sound idea to clear out. No one could sleep in peace and it was inevitable that some one would be bitten eventually. So we struck the tents one bright September morning and our cars roared down the hill leaving Viper Camp to the snakes and vultures.

During all the time that I was in Borneo and the East Indies I met but very few snakes. The reason of course is because the jungle is so dense that you just don't see them. They are there all right but they don't advertise themselves. I always had the uncomfortable feeling that I might sit on one, for some of them are so protectively colored that they are almost invisible. Particularly was I nervous after an experience I had with a huge python which might easily have cost me my life.

I was walking up a game trail in the dense jungle with my Filipino boy just behind me. His name was Miranda. Suddenly he grabbed me by the shoulder and jerked me violently backward.

"What the devil," I began.

"A snake, Master—a big snake. There—right in front of you—on that tree. You shoot him quick!"

He pointed to a branch overhanging the game trail only a few yards away. I couldn't see anything. Miranda kept pointing to the tree in front of me, hopping up and down with excitement.

"There, Master—can't you see it—a big, big snake."

No, I couldn't see it. Suddenly a breeze stirred the palm leaves and a spot of sunlight drifted over the branch where Miranda was pointing. Then I saw. There lay a huge, ugly flat head and a black glittering eye. Following back from the head what I thought was a tree branch developed into the snake's body. There seemed to be yards and yards and yards of it, losing itself in the thickness of the tangled vines. I backed away thirty or forty feet and lined my sights on that glittering eye. At the crash of the rifle, a typhoon seemed to have been let loose in the jungle. I turned and ran, watching from a safe distance. The writhing coils of the gigantic serpent broke trees like pipe stems, sweeping away in slashing blows everything within a dozen yards.

Half an hour passed before the reptile was still. There in the jungle cleared by its own death struggles lay the snake, its body half as thick as my own. I thrust a stick between one of the coils and even though life was gone they tightened like a vise. Miranda and I straightened out the great reptile and paced its length. It measured twenty feet.

As I looked down at it I thought what an unpleasant death I might have had except for Miranda's sharp eyes. It was lying there alert but motionless, waiting for whatever living thing moved along the trail. A wild pig, a deer—or myself—anything that passed under the branch would have been swept into those terrible coils and crushed to death.

That experience did not enhance my love of snakes. Neither did another which I had in a *dâk* bungalow in Burma. It had been weeks since the last traveler had stayed there, and when it was opened, I told the native caretaker to make a thorough search for snakes. A good thought it was too, for in the bathroom behind the tin tub a cobra was placidly waiting for rats or what else might chance to pass his way. He wasn't a very big cobra but quite large enough to have sent me to the Happy Hunting Grounds if he had taken a nip at my bare leg. A cobra is particularly bad medicine for he doesn't just strike and pull away like most other respectable snakes. He sinks his fangs and hangs on, letting the poison drain into the wound till the last drop. The fact that the cobra is such a big snake and consequently has so much poison accounts for the fact that the chance of recovery from an honest-to-goodness bite is pretty small.

My last snake story is about a ship and a quartermaster. The ship was the United States deep sea exploring vessel, the *Albatross,* and the quartermaster was a Swede by the name of Larson. Larson was a fine sailorman but he would get drunk. He had been on the *Albatross* for twenty-five years—ever since the day that she was launched. I remember how when a blow was coming on, he used to walk about the vessel seeing that everything was tight, saying:

"Now we are at peace. Now we are at peace."

God knows I didn't feel much at peace at those times —I was too seasick! We had been at sea for some weeks when one morning Larson turned up drunk as an owl.

He had a great bearded face, which, when he was drunk, looked just like that of an amiable orang-utan. There were a lot of young boys on the ship who had enlisted in the navy to see the world. Larson would select one of these kids, thrust his great face within an inch of the boy's nose and with a horrible grimace say "boo." He would roar with laughter when the kid nearly went over backward.

The first morning that Larson reported for duty drunk, the captain ordered him to the brig until he sobered up. But only a few days later Larson was drunk again. This happened half a dozen times and even the crew became very curious to know where he got his stuff. The ship was searched but it remained a mystery until I solved it by chance.

While we were in the East Indies I had collected several five-gallon jars of snakes and lizards, which were preserved in alcohol. These were in the storage room below the laboratory. Every few weeks I looked them over to renew the alcohol which would become diluted by the blood and fluids of the animals' bodies. I discovered that the jars were almost dry and that this was the source of Larson's jags.

The old fellow had reached the point where he just had to have something to drink and the snakes and lizards offered the only solution. He confessed to me afterward that it did taste pretty awful but that the effect was swell.

CHAPTER IV

Random Reminiscences of Animals

I OUGHT to have had many narrow escapes from wild animals. The trouble is that they really *are* wild and they want to be let alone. Unless one is wounded or cornered or has babies, even the most dangerous animals will usually run at the sight of a man or the sound of a human voice.

I have shot literally thousands of animals. It has been a part of my job as a scientific explorer for a great museum that must have specimens for study and exhibition. Yet only one unwounded animal ever deliberately charged me. That was a wild boar in Korea when the beast was driven out of cover by beaters. He came for me through the tall grass with foam-flecked jaws, his little eyes flaming red. I dropped him only a few feet from the end of my rifle.

Of course dozens of my sportsman friends have had narrow escapes from charging animals but it was only because they picked a fight. All the wild cattle anywhere in the world are dangerous. They may come for you on sight. Lions seldom will, or tigers either. Elephants are an uncertain quantity and so are leopards.

My most interesting tiger experience I have told before. Still I think it will bear repeating, for it shows how even a man-eater will prefer to clear out before a human even though he be comfortably settled in his lair.

It was when I was in Korea, and the tiger was the Manchurian variety. In length he really isn't larger than the Indian species, but the inch-long hair makes him look a lot bigger. Since a good part of his life is spent in the snow, he is much lighter colored than the one in the south. There aren't many white men who have killed either the Manchurian or Siberian tiger. There was just one individual in that particular region of northern Korea—or at least one that we knew about. He was a big fellow—a man-eater—and the natives were terrified whenever he made one of his visitations.

His habit was to go from village to village on a certain beat, picking up pigs, chickens, dogs, and now and then a child. It may be true that once a tiger tastes human flesh he likes it so much that he refuses all other nourishment, but I doubt it. Anyway, it isn't true of the Chinese tigers nor was it of my friend up there in Korea. He much preferred a mangy dog to a fat Korean child. I was told by a reformed cannibal on an island off the coast of New Guinea with whom I had a heart-to-heart talk about the flavors of various members of the human family that white men tasted much too salty—that he infinitely preferred natives who didn't have such a strong saline flavor. Perhaps my tiger felt the same way about it. Be that as it may, he had killed lots of dogs and pigs and a few children. I was keen to rid the community of the big cat, but apparently he did not share my enthusiasm to effect his early demise. He never stayed long enough in any one place for me to catch up with him. A breathless native would arrive at my camp saying that the tiger had been seen at his village twenty miles away. In fifteen minutes I'd be gone with my hunter Paik *sontair.* Paik was a fine old Korean who had killed three tigers with his flint-lock gun. He had been given the title of *sontair,* which was equivalent to a knighthood. We would arrive at the village where our tiger had last been seen, only to draw a blank. In a day or two news would come that he had appeared somewhere else.

We tried all sorts of ways to achieve a personal acquaintance. Sometimes we cut across his line of march, hoping to be there when he arrived. Then he'd double back on his trail and appear again at the village which we had just left. I tried just waiting—sitting in one place until he came. Nothing doing! My shoes were worn thin from traveling, my patience was exhausted and I had just about decided that he really was the Great Invisible. Then one day a native arrived in camp saying that he had seen the tiger go into a cave near the top of a mountain only a few miles away. Paik and I were there in less than an hour. In the dust at the entrance to a black hole in the rock we saw the fresh pug marks. All led in and none came out. We were sure we had him that time.

We concealed ourselves in a clump of bushes a few feet from the cave mouth and sat down to wait. I was tense with excitement but my taut nerves began to relax as hour after hour passed and nothing happened. The sun went down and it grew dark. Still there was starlight and the tiger couldn't get out of that cave unseen.

It was a long night and in the first gray light of

dawn we examined the dirt at the cave mouth. No new tracks.

"If he won't come out, we'll go in and get him," announced Paik in the most matter-of-fact voice. Just like that! "We'll go in and get him."

I hadn't lost any tigers in that cave. Besides I needed food; my tummy was awfully empty. I had been up all night—I was tired—and most of all I was scared green at the thought of crawling into that cave with the tiger sitting there comfortably ready to receive us.

When Paik said, "Are you going?" I gave him an evasive answer. More explicitly, I told him to go to hell.

He stared. "You've got a light," said he. "The tiger won't charge the fire. He'll be frightened."

Well, I was frightened too—awfully frightened.

"Besides, I'll be right behind you with my spear," announced Paik as though that made everything all right. (I forgot to say that Paik carried a spear but no gun.)

I gave him a still more evasive answer. Then he lost all patience.

"If you are afraid, let me take your rifle and I'll go in," he said, and the look he gave me told just what he thought about white men who pretended they wanted to kill a tiger.

That was a bit too much. To have that old native calmly tell me that he'd do it alone if I were afraid brought forth all the fighting spirit of my ancestors. I thought of George Washington and I knew he'd have walked right into that tiger's cave with his little hatchet. He wouldn't even have hesitated, and here I was with a high power rifle refusing to budge.

I said to Paik, "Let's go."

We had to crawl on hands and knees, for the cave was less than shoulder-high. I went first, rifle in one hand, electric flash in the other. Paik followed with spear advanced. As a matter of fact, it was advanced so much that it kept pricking me in the rear end to my intense annoyance. I wouldn't swear that he did it on purpose, but I suspect him strongly.

About twenty feet from the entrance there was a small chamber. A sickening smell of rotting meat almost suffocated me, but no tiger. We could see that the passage turned sharply to the left. I felt morally certain that that infernal tiger was waiting just there ready to reach out and claw me when I turned the corner. Paik seemed to know that I had lost my nerve and gave a few extra prods with his spear. With my flash-light stuck way out in front, I slowly edged around the rock. Thank God, no tiger. The passage led on, dipping slightly downward. In the distance I saw the faint gray of daylight and knew that the tiger wasn't there.

We emerged on the other side of the peak in a deep gorge filled with huge boulders which concealed the mouth of the cave. Pug marks plainly showed in the soft sand. They were leading outward. The natives of the village did not know that there was an exit to the passage for they had such fear of tigers that no one would go near that part of the mountain. Our animal evidently had scented us at some time during the long hours that we were watching the entrance of the cave and had quietly slipped out the back door. We heard of him again next day at a village twenty miles to the south.

Random Reminiscences of Animals

In America I suppose that the grizzly bear is our most dangerous animal. I've never had any personal experience with them for I've done but little shooting in the United States. But some years ago I hunted Alaskan brown bears on Kodiak Island. These are the largest living carnivores. A big one will weigh nearly fifteen hundred pounds. In the spring when they have just emerged from their long winter sleep they are hungry and pretty nasty customers to meet on your evening walk. There isn't much for them to eat except roots. A guide told me that he had seen a bear in the spring rear up on his hind legs and strike and slash at a tree just out of ill nature. He will come for a man at sight then. There are plenty of well authenticated instances of men being killed or badly mauled by these huge beasts, though as a rule a bear isn't a very dangerous animal.

In the autumn, when the Kodiak bears are rolling in fat preparatory to their long hibernation, they are too lazy to attack unless really cornered or provoked. They much prefer to amble away, though they aren't going to rush off and be undignified about it.

I was on Kodiak Island when the salmon were running. Some of the streams were so choked that the fish could hardly move. It was just like the proverbial sardine tin. These salmon work their way up the rivers as far as they can, lay their eggs and then die. Literally millions of them come in from the sea and it is a time of feasting for the bears, who love fish.

I was following up a wide stream bed. At that time the water was running in a channel on one side, but in previous years it had followed other courses. Just in front of me was an old log jam—a chaotic mass of trees and branches lodged there when the river was in flood. Picking my way carefully up the jam I had almost reached the top when I saw another jam about a hundred yards away. Just then the head and neck of a bear appeared above the rampart. An enormous head it was too—perfectly huge. I was shooting my old 6.5 mm. Mannlicher and lined the sights on that great neck. I thought the little bullet would smash the cervical vertebræ and there would be no more argument with that bear. At the crash of the rifle the head disappeared. Just as I threw in another shell the head bobbed up again twenty feet away.

"What the devil is the matter with me?" I thought. "I couldn't have missed at that range."

The second time I rested my rifle against an upright stub and held just under the great muzzle. When I fired down went the head.

Throwing in another shell I climbed to the top of the jam. There, by Jove, was my bear running away. I took a quick shot and the beast dropped. Up again but at the second shot it was down for good. When I reached the bear I found it to be an enormous female, very fat. I looked her over and discovered that the first bullet had gone through the lungs; the other had smashed the back.

I was curious to know if I had hit her with the first two shots. Not a sign of a bullet anywhere in the neck however. I straightened up wondering how I was going to skin the huge beast, and looked about. To my intense surprise there lay a second bear at the foot of the log jam and twenty feet away still another.

It was a mother with two full grown cubs nearly as

large as herself. My first two shots had been absolute bull's eyes. In both cases the neck vertebræ were smashed to pulp and the bears had died instantly. Later I wounded another bear which got into a thick jungle of alders. I wanted to follow up the trail where we had to go mostly on hands and knees, but the guide would have none of it. He said that these animals when wounded have a habit of lying down in the alders and charging out like a hurricane. He told me of three instances of hunters being killed in this way.

I suppose that the narrowest escape I ever had from being killed by an animal was during my early days of whale hunting. Whaling as conducted from the little modern steamers doesn't begin to have the danger of the old deep-sea method when the animals were harpooned from small boats. Still, there is plenty of excitement always, and now and then one has a real adventure.

One bright summer's day while hunting whales off the coast of Japan we got fast to a big finback seventy feet long. The harpoon struck him between the shoulders and as the bomb didn't explode he was virtually uninjured. Dashing off like a hooked trout, he took out rope so fast that the brakes on the winch were smoking. Cable after cable was spliced together and before his rush could be checked he had out nearly a mile of line. Then the brakes were set and he towed the ship forward with engines going full speed astern. After an hour of this even his great strength began to fail. The rope was slowly reeled in but we could never get closer than about half a mile. Then a wild dash would take him off again in a smother of foam. The fight dragged for six hours.

"He'll keep us fast all night," said the Captain. "I'm fed up with this. I'll send a boat out and lance him."

"Let me go," I asked. "I want to get some close-up photographs. I'll pull one of the oars."

"All right. You'll take the *praam*."

A *praam* is a Norwegian boat big enough for three or four men which sits deep at the stern and can be spun around almost like a top.

The mate took a long slender lance. A seaman and myself were at the oars.

The whale lay at the surface nearly a mile away, now and then blowing lazily. As the tiny boat slipped up from behind, the body loomed bigger and bigger to my excited eyes until it seemed like a half-submerged submarine. Standing in the stern with lance poised, the mate steered us up right beside the whale.

"Way enough," he whispered.

Swinging the *praam* about, we backed up till the boat actually touched the gray body. Bracing himself he plunged the slender steel deep into the animal's lungs.

As his arm went down we gave a great heave on the oars. My right oar snapped short off and the *praam* swung directly against the whale. Up went the great body. I saw the tail, twenty feet across, weighing more than a ton, waving just above my head. It appeared to hang in mid-air and then to be coming down right on me. Never will I forget that second—it seemed hours long. The tip of the fluke missed me by six inches but the side of the boat was smashed like paper.

I was in the water with the other two men swimming for the floating wreckage. We caught hold of the stove boat and looked around. The whale lay at the surface a few fathoms away, blood welling out of the blow holes. We could hear the rattling of the winch as it wound in the rope while the ship crept nearer.

Suddenly I felt something bump my foot. It was a huge shark. The water was alive with flashing white bellies and great sharp fins. I was yelling like mad and absolutely sick with fright. The other two men joined the chorus. They were as scared as I. Wrenching off pieces of wood from the broken side of the boat, each of us hung on with one hand, beating the water with the other. But the sharks weren't interested in us. They swarmed like flies about the dying whale, drawn by the blood pouring out of its nostrils.

The ship came alongside us and the Captain shouted, "Hang on a little longer. I want to kill this whale before I lower a boat."

With that he calmly left us and went over to the dying finback. We yelled and cursed while the whale was drawn up under the bow and lanced. The Captain only grinned. When we were on deck again I walked up to him as mad as a hornet.

"What the devil did you mean by leaving us there among the sharks while you killed that damned whale?"

"Well, you were all right, weren't you? The water is warm, you had a nice boat to hang onto and as for those sharks—shucks, they aren't man-eaters."

There's just no use arguing with a man like that.

Of all the whales I studied the killer whale and the California gray whale or devilfish were the most interesting. The latter was particularly so because for forty years it had been supposed to be extinct. I rediscovered it off the coast of Korea.

Devilfish, they are called, because they fight like the devil when harpooned, but all the fight goes out of them when the killer whales arrive. Those babies, armed with a double row of teeth, literally devour the devilfish alive in spite of the fact that the devilfish are nearly fifty feet long and the killers only half that length. What the killers particularly like is the tongue—and the living tongue at that. I actually saw a killer come up beside a devilfish, stick his nose up against the lips, put on full speed with his tail and force his way into the devilfish's mouth. The whale was paralyzed with fright and lay there on its back, flippers wide spread while the killer devoured great chunks of its tongue. In the meantime, other killers were tearing at the throat and belly as the poor beast rolled in agony. We shot two devilfish one day and made them fast to either side of the ship's bow. As we towed them in to the station a killer swam alongside the ship, forced his head into the half open mouth of the dead whale, and started work on the tongue.

Killers are called the "wolves of the sea," and it's a pretty good name. They hunt in packs and will devour anything that swims. Captain Scott writes in his diary about a narrow escape that Ponting, his photographer, had from killers in the Antarctic. The ship was moored to an ice floe and Scott called to Ponting to get some photographs of a school of killers which were swimming along the edge of the ice. Apparently they were interested in several dogs which were on the floe. Ponting ran out just as the killers came up under the ice, smashing it with blows from their great bodies. Ponting barely escaped falling into the water. The ice which the killers broke was three feet thick. Admiral Byrd, too, had a narrow escape from killers in the same place.

To my mind a killer whale was responsible for the story of the "Loch Ness Monster" in Scotland which made headlines in the public press a year ago. For some reason the man-in-the-street likes to think that sea serpents do exist; that somewhere in the ocean depths or in some far-off sea there may be strange creatures, survivors of the Age of Reptiles, still living. In spite of the fact that there isn't a shred of scientific evidence in support of such a belief it will not down. It has existed for centuries and I suppose will continue to exist for more centuries to come. All the sea serpents that have been reported, if they ever were captured have proved to be some already known fish or animal.

Faulty observation is responsible in most cases. An animal at sea, as a rule, gives only a brief glimpse of itself. Perhaps it is a whale or a shark or a seal which is unknown to the observer. He sees it for a second or two, and only part of the body at one time. If he has a lively imagination he can make it look like almost anything—and really believe that he saw what he thinks he saw. A sea serpent, to fit the popular conception, should have a long neck and a head like a camel. Why a camel, God only knows, but a camel-like head it must have.

Our observer tells of his apparition to a news hawk. A sea serpent story is just duck soup to a reporter. The story appears with a few added embellishments. Then the game is on. All over the country, along every seacoast—even in little frequented ponds and lakes—sea serpents will begin to appear. And they will all have a head like a camel! You watch for the next one and see if I am not right. Heated arguments with newspaper men have been my lot when I have scoffed at the sea serpent myth. The last one was with the editor of a great New York paper.

"What I can't understand," said he, "is the pig-headedness of you scientific fellows. Just because you haven't seen it you think it can't exist. You make me tired. You ought to put a sign up on your Museum 'Abandon all individual opinions ye who enter here.'"

Arguments are of no avail. The average person wants to believe in sea serpents and he or she is going to believe in them, no matter what facts are presented to the contrary. I suppose they feel as I felt as a child when other children tried to make me believe that there wasn't any Santa Claus. Newspaper men like sea serpents, of course, because they give good copy. A story like that can be kept alive for days.

I've said that most of the stories are the result of faulty observation. I'll explain what I mean. When we were hunting devilfish off the Korean coast, Captain Melsom and I saw a whale swimming directly across our bows, half a mile away. It had two dorsal fins. Now no whale known to science has two dorsal fins. None at all, or one, yes; but not two! Nevertheless there it was. Melsom saw it, I saw it and all the crew saw it. There was not a man on the ship who would not have sworn an affidavit that we had seen a whale with two dorsal fins.

I said to Melsom, "We've just got to kill that whale if we follow him a week. It's something new to science. If you kill it I'll name it after you."

The whale was loafing lazily along at the surface, apparently not going anywhere in particular. Melsom changed course to come up astern of the whale and we slid along quietly, engines at half speed. We were pretty close and the dorsal fins began to look rather strange. They weren't in line as they ought to have been. Then suddenly the whale dived. A great tail hove out of the water and right beside it a smaller tail. The mystery was solved. It was a mother finback whale with a half grown calf pressed close to her side. The calf was in such a position that its dorsal fin appeared just behind that of its mother. As we first saw them, with the calf on the far side and completely invisible, it looked exactly as though the mother had two dorsal fins. If we had not followed it up, I most certainly should have announced that I had seen a double-finned whale. It would have gone into scientific literature and remained as a recorded fact, backed by the statements of the other men on the ship. That's what I mean by faulty or incomplete observation.

The Loch Ness Monster was seen several times and photographed once. The picture, as it appeared in the New York *Times,* showed a long curved neck surmounted by a small head. Evidently it had been retouched by an artist who was obsessed by the "camel head" complex. I got a copy of the original from the *Times* and it showed just what I expected—the dorsal fin of a killer whale. A killer's dorsal is six feet high and curved. It would make a wonderful neck for a sea serpent. The head could easily be supplied out of the imagination as the newspaper artist did, in fact. Doubtless what happened was that the killer made its way through the narrow gate of the Loch from the open sea and remained there for some days. It may even have gone in or out several times.

But wasn't I unpopular with the Loch Ness supporters when I said what I believed the "Serpent" to be! As a matter of fact the mythical creature was a boon to the dwellers near the Loch. Tourists came there in thousands and the canny Scotsmen did their stuff. No tourist got away without leaving a certain amount of "siller" to help keep the wolf from the door. I have a collection of picture postcards purchased at Loch Ness and the "monster" is shown in half a dozen forms, none of them alike.

Whales and sea serpents seem to have risen up lately as a ghost out of my past. Sea serpents because of the Loch Ness Monster and whales because of the Dionne quintuplets. When I read of the arrival of the famous sisters way up there in Canada I hadn't an idea that I should ever figure with them in the public prints, even in the remotest way. But one day last winter Doctor Dafoe, the quintuplets' "deliverer," arrived in New York. The keys of the city were virtually handed over to the little doctor. New York set out to give him a good time and if he didn't have it at least the city did. He went everywhere and saw everything from Al Smith atop the Empire State Building to a ride on a ferryboat—his first, by the way.

It was reported in the papers that the doctor had said that the two things he most wanted to see in New York were the New York Zoölogical Park and The American Museum of Natural History. In my capacity as Director of the latter institution, I extended an invitation for him to visit the Museum. He arrived one morning accompanied by his brother, also a doctor, two detectives and half a dozen newspaper and cameramen. I met them at the entrance.

"You haven't much time, Doctor," said I. "What would you most like to see?"

"Whales," he answered without a moment's hesitation. "I read a book of yours on whales and I want to talk to you about how they are born."

The words were hardly out of his mouth before the news hawks were on the job. Oh boy, what a story! Quintuplets and whales! But the doctor was really serious. He had never thought that it would make a good story. He isn't that sort of a doctor. By some chance he had come across a little book which I had written years ago entitled "Whale Hunting with Gun and Camera." In it I had written quite a lot about how whales are born and nursed and reared by their whale mothers and it had made him very curious.

We walked out to the Hall of Ocean Life which is mostly filled with specimens collected by me during the eight years which I spent sailing the seven seas in search of cetaceans. Not only did I collect whales for the Museum during those years, but I developed an almost indecent tendency to pry into their private lives. I was a regular Walter Winchell. Nothing was safe from my field glasses and camera and notebook. Probably a whole series of whale divorces could have been started if it had become public in the whale world what I saw and photographed when certain lady and gentlemen whales were stepping out with each other on the surface. I saw them make love, and play together, and was even present at the *accouchement* of a whale, albeit it occurred when the animal was dead and was being drawn out of the water by a steam winch. Still I was present at the birth and I could give Doctor Dafoe a lot of first-hand information as to the way it happened.

"First," said the doctor, "I want to know how big is a whale at birth?"

"Well, that depends upon the species and the size of the mother, of course. I took a baby thirty feet long, weighing eight tons out of an eighty-foot blue whale."

"My goodness! Eight tons, did you say? That's sixteen thousand pounds. My goodness! I never thought they were as big as that. What an awful time the mother must have. Why are they so big, do you suppose?"

"It is because they live in water, which is a supporting medium. Of course if the mother had to carry such a weight about on land, she couldn't do it."

"I read in your book, Doctor Andrews, that you had actually tasted whale's milk. What does it taste like and how does the baby nurse?"

"The milk I tasted was pretty strong—a decidedly fishy

taste—but it looks just like cow's milk. How they nurse, though, I never could figure out. The two teats are only about two inches long and the baby whale has a pointed snout and great thick lips. I suppose the mother rolls over on her back with the teat out of the water and the milk is ejected into the baby's mouth. I can't see any other way to do it. Certainly the baby couldn't nurse under water."

All this time we had been walking about the hall. I tried to show the doctor the wonderful coral reef group. He looked at it interestedly but always kept coming back to the whales.

"Did you ever see a whale actually making love?" he asked.

"Yes, once. I suppose I am the only scientist who ever did."

"What was it like?"

"Well, in the main, it was just like any land mammal. For, of course, whales are land mammals that have taken up a life in the water. They would drown if they stayed under the surface too long. They used to live on land millions of years ago. The hind legs still remain as rudiments. The pair of whales that I saw making love were humpbacks fifty feet long. It was in Frederick Sound, Alaska. There wasn't a ripple on the surface and the spouts of the two whales shot up like a cloud of silver in the sunlight. Then one of them literally stood on its head and began to wave its great tail in the air, pounding the water into white foam. The other whale lay at the surface, rolling about indifferently. Then the bull swam up alongside and gave her a love pat with his twenty-foot flipper that would have killed an elephant. For ten minutes they lay still; then the bull backed off and dove. I thought he had gone for good, but she wasn't worried; she knew that he wouldn't leave her—yet. Suddenly, with a tremendous rush the bull threw himself completely out of the water like a leaping salmon. It was a magnificent sight as that enormous body, weighing nearly fifty tons, shot into the air. He fell back with a mighty splash and lay still. Then, rolling over and over like a log, he slid up to the female and clasped her with his great flippers.

"All this time the ship had been creeping nearer. We were only twenty fathoms away and from the barrel at the masthead I could see every detail. Captain Grahame stood at the harpoon gun ready to shoot."

"I yelled to him not to kill them then, but he was thinking of the thousand dollars their great bodies would bring the company. He shook his head, sighted along the barrel of the gun and let drive. The bull was dead instantly and the female still lay at the surface apparently dazed. Another harpoon was rammed home and the captain sent it crashing into her body just as she started to dive."

The doctor was listening, fascinated, to the story.

"My goodness, I'd like to have been there. Did you see everything?"

"Yes," said I. "I saw everything."

We wandered out of the Hall of Ocean Life to the Roosevelt Memorial, still talking about whales. Of course I asked him about the famous quintuplets.

"Well," said he, "I don't deserve any particular credit

for just delivering them. That wasn't so difficult. But keeping them alive wasn't so easy in all that dirt and in the unsanitary conditions. The organizing afterward—that was the real job. And then, too, there were religious difficulties. But I've been there a long time and they sort of trust me."

With a deprecating laugh he dismissed the matter. A fine type, the doctor. One of those men who radiates dependability. He wasn't swept off his feet by the publicity and the reception New York City had given him, either. He knew just what value to put upon it. He was enjoying it thoroughly but his wise gray head had not been turned one little bit. A simple, kindly man devoted to his work and the people among whom he lives. I don't wonder that they "sort of trust him."

CHAPTER V

Wolf of Mongolia

EXPLORATION and dogs are inseparable. At the Explorers Club in New York City we have one wall devoted to the photographs of famous dogs who have made honorable names for themselves in exploration.

I had two dogs on the Central Asiatic Expeditions. The first was a Siberian sledge dog named Mushka; the other an Alsatian police dog, Wolf. I might easily write a chapter about Mushka, for he was an extraordinary animal. But he never captured my affection as did Wolf. Mushka was the most egocentric animal I have ever known. He thought only about himself and acknowledged no man as master. He never showed a strong attachment to any individual. The person who would give him food when he wanted it and let him indulge his passion for hunting was the one to whom he rendered temporary allegiance. But it was only temporary. If he decided that some one else would give him more hunting or better food he dropped his other friend like a hot coal. He had uncanny intelligence, but it was exercised only for his own benefit. He used to move about in his little world completely wrapped in a mantle of selfishness. I couldn't help admiring him for his competence, but I liked him no better than I like a man or a woman with similar characteristics.

Wolf of Mongolia

I bought him as a hunting dog from a friend of mine, Oscar Mamen, in Urga, Mongolia. My first experience with Mushka was rather humiliating for me. Before I purchased him I borrowed him from Mamen to hunt roebuck. We were in the heavy forest north of Urga when a fine buck jumped out of the scrub birch. Mushka was off like a bullet. Not a yelp or bark—just a soundless red streak on the trail of that deer.

Not knowing Mushka's method of hunting, I followed. That was my mistake. I should have waited just where I was. In ten minutes the big red dog had brought the deer around in a circle to the place where he had started. I wasn't there. Mushka dropped the hunt instantly and took up my trail. He came up and stared at me. If ever an animal's face showed utter disgust and contempt, his did. After he had impressed me with what a poor fish he thought I was, he turned about and went straight back to Urga, without a glance in my direction. He wasn't going to waste his valuable time in working with a guy like me. Later, however, after Mamen had explained his method to me, I killed a lot of deer for him and he accorded me the privilege of his companionship in the field.

Wolf was a different type. Loyal, affectionate, intelligent, courageous—he was everything that a dog should be. He had been "mentioned in dispatches" a score of times when we returned to Peking from the Gobi. I loved him and it tore my heart to leave him in Peking. He died only a few months ago. The Associated Press and United Press sent his obituary throughout the world, for he was a real explorer. A short time before his death I wrote the following article. It seems fitting to include it as a chapter of this book just as it was written, for I can never think of Wolf as dead.

Wolf pattered over the stones of the courtyard into my bedroom in Peking. He gave me a judicial look, then gently nudged my arm with his cold nose. He had decided that it was time I went to sleep. We understood each other perfectly. Yes, it was midnight. When the light went off Wolf curled up in his big chair. He always slept there except now and then when he stole into the drawing-room to stretch out luxuriously on one of the brocade sofas which he knew right well were forbidden to all dogs.

I had been asleep for three hours when suddenly Wolf gave that peculiar growling bark which meant only a man-hunt. He was out of the door like a hurtling black demon. Came a scramble and a stifled scream. By that time I was on the floor, revolver in hand. Wolf was circling about the ten-foot compound wall, every hair on end, snarling throatily. Together we went through the rock garden and among the trees and bushes, but he always returned to the wall. Obviously the intruder had escaped the way he came and I knew that Wolf would sleep no more that night, so I went back to bed. The next morning Lo, my Number One Boy, and I made an inspection. Below the wall there was the imprint of two feet and a strip of blood-stained cloth. Evidently it came from a pair of Chinese trousers. Lo looked at it and grinned.

"Wolf take plenty meat out of that Chinaman," said he.

No doubt about it. The unfortunate thief had made just two jumps—one down and the other up. If there had not been a missing brick in the wall that gave a toe-hold he would have been a dead Chinese. Wolf doesn't waste time on externals. When he attacks he goes straight to the root of the matter—the throat. He has been taught that by countless fights with the man-eating dogs of Mongolia. Had he caught the thief even if I could have pulled him loose, which is doubtful, the police would have given the man short shrift the next day.

This, of course, because I am a foreigner who has extra-territorial rights in China. But a Russian friend of mine, Friedlander by name, had a very different experience. Since the white Russians have no national representatives in China, the native police and other officials lose no opportunity to humiliate them. Friedlander's house was robbed one night when he was away and several thousand dollars' worth of furnishings stolen. The next day he brought a police dog from his house in the country. A week later a second robbery was attempted but this time the police dog caught the thief. Before his master could pull him off, the dog had almost undressed the man and badly chewed his left leg. Friedlander turned him over to the police and the thief confessed that it was he who had robbed the house the first time. He had returned to complete the job.

Instead of dealing with him as would have been done in the case of a foreigner with extra-territorial rights, the Chinese police saw a heaven-sent opportunity for "squeeze." They told Friedlander that he had no right to keep such a fierce dog in his compound. That he must pay a thousand dollars fine for the injuries to the thief and in addition stand his hospital expenses. All because he kept a watch dog in his own house!

Wolf always slept with both ears open. He must have, else he could not have heard a robber who tried to loot the Expedition's equipment room. This was a separate house, backed up against the street wall a good hundred yards from the rear court where I slept. One morning we found that the door had been forced and two new automobile tires were lying at the foot of the compound wall. Wolf had heard the thieves and frightened them off as they were getting over the wall just as he did the one near my bedroom.

It is hardly fair to introduce Wolf as a man-hunter, for he is the most affectionate, the gentlest and the most lovable of police dogs. But the responsibility of the house and the camp rests on his shoulders and he takes it seriously. If a visitor arrives *via* the front gate, properly chaperoned by the *kan-mundi* ("look-see-door-man") Wolf meets him with delighted barks. He loves people and loves to be petted. But, God help the intruder who comes in over the wall or in any other unorthodox fashion. Then Wolf is a raging man-killer.

Eight years ago he was born in Kalgan, that little frontier city on the edge of Mongolia, with five other pups. Jack Strange loved him but he never could keep him at home. Always Wolf was ranging the hills. Sometimes he would disappear for a week. He looks so much like a wolf that Strange feared he would be shot by wandering Chinese soldiers. Thus he came to me. Norman Lovell, one of the Expedition's motor transport officers, brought him to Peking on a cold winter's night. There were four of us in the office when he arrived. Wolf went from man to man, sniffing each one and regarding him gravely. Perhaps it was because I gave him the back of my hand to smell while the others presented their palms. Anyway he flopped down on the floor with his head against my feet. When we left the room an hour later I was the one he followed. He had adopted me for better or for worse. He was fond of all the other members of the Expedition, but his allegiance to me never wavered.

He loved camp life in the desert. As soon as preparations began in the spring Wolf was bursting with excitement. He knew perfectly well what it was all about. One April we sent seven of the motor cars across the city in the evening to be loaded on the train early in the morning. Only one truck remained in the compound. I was ready to go at daylight but no Wolf. We searched every corner of the gardens and every room of the house. Not a trace. I couldn't wait any longer and left orders to have a coolie bring him to Kalgan by the next train. A tarpaulin covered the truck, tied down by ropes except in front. I was driving and was halfway to the city gate when a cold nose touched me on the back of the neck. Wolf's head was cautiously extended from under the tarpaulin. His look was half pleading, half reproach. I stopped the car and put my arms around his neck; with a yelp of delight he struggled out onto the seat at my side. Poor old dog! When he saw the seven other cars driven off the previous night he thought we were going to leave him behind. I never could figure out how he had managed to burrow under the tarpaulin, for the ropes were pulled tight and he is a large animal.

Wolf had come to believe that all dogs are his enemies. It is natural enough because he is a foreigner in Mongolia and the big Mongol dogs go for him like a shot. They seem to object to the fact that he is different. He certainly is a lot more handsome, and perhaps they resent his personal beauty. Anyway, there is no sniffing around—they fight at sight. Wolf has learned that attack is the best defense and he generally beats them to it. What fights those are! Wolf has one advantage. He wears a broad spiked collar and that protects his throat. Also, he is cleverer than the others. But they have thick matted hair and sometimes it is difficult for him to get a throat-hold. I have seen him spit out three or four mouthfuls of hair before he could get his teeth in flesh.

Wolf has discretion as well as valor. He is perfectly willing to take on any two dogs in Mongolia even though they may be bigger and heavier than he, but he watches his step when there are three or more. The Mongol dogs are terrifying beasts. Larger than a police dog, black with brown points, and savage as tigers. From earliest birth they are taught to guard the caravan or camp and will attack a stranger at sight. Not only that, they will eat him too. The Mongols do not bury their dead. Instead, they throw out the corpses to be devoured by the dogs, wolves and birds. Near a lama monastery packs of dogs that live largely upon human flesh are always prowling about. In Urga, the capital of Mongolia, it is most unsafe to go out at night unarmed. I came into Urga one evening on horseback after a seventy-mile ride. My pony was dead-tired and trotted slowly up the main street. Suddenly five large dogs rushed from behind the old Russian Consulate. One leaped for my leg, but I kicked him off and he caught the stirrup leather; another fastened his teeth in the pony's hind leg and a third got a hold on his tail. As I pulled my rifle from the holster the pony lashed out wildly with both feet. One dog rolled over yelping and instantly the whole pack was upon him. I watched a pack of dogs tear apart a dead Mongol who had been dragged out from the Lama city. It took just seven minutes to scatter the corpse over the plain. I won't give you the details for it wasn't a nice sight.

I had a narrow escape myself from being eaten by fourteen dogs while lying asleep in a fur bag on the desert near the Turin monastery. Only a lucky shot from a tiny .22-caliber rifle which killed the leading dog and turned the pack saved me from a horrible death. Almost every member of the Central Asiatic Expedition has been attacked and we made it a rule never to move from camp without a revolver.

Thus you can see what pleasant companions Wolf had in Mongolia and why he doesn't like them. He is a dog of breeding and character and it is a long way beneath his dignity to associate with canines that eat human flesh and are cannibals besides. He knew that he never would get a square deal; that they would turn on him at the first yelp, so he declared war on all dogs.

Wolves he doesn't like, but I think he respects them. He knows they are bad medicine even for a fighting police dog. The Mongolian plains wolves are long-legged rangy beasts perhaps a little heavier than Wolf. He will bark at them from a safe distance, or if one of us backs him up with a rifle, but he won't attack in earnest. Two wolves came near camp early one morning. Wolf routed me out of the tent by frantic barking. When he saw that I had my rifle and was following, he dashed furiously at the wolves as though he would finish them both off. They loped slowly away at first but suddenly stopped and faced him. You should have seen him put on brakes with all four feet! He wasn't having any, thank you. As he saw me drop on one knee to shoot, he jumped to one side and waited for what he knew would come. I killed the nearest one but the other was only wounded. He whirled about and tried to run but Wolf was on him like a tiger. When I came up he had his deadly throathold and the story was almost ended. Again early one morning a wolf walked right into camp; I suppose it was curiosity. We tried to "sic" Wolf on the animal but he wouldn't be "sicked." He gazed about the landscape in the most interested manner in every direction except at the wolf. He knew that we were not backing him up with a rifle and he didn't intend to be the goat for anybody's fun.

Gazelle drove him wild for he couldn't catch them. One morning a great herd came up out of the badlands not a hundred yards from the tents. They streamed over the rim of the basin in a yellow flood, thousands of them. Wolf would dive for an antelope like a bullet. It would wait until he was almost ready to spring, and then leap away. Since a gazelle can reach a speed of sixty miles an hour for the first dash, poor Wolf didn't have a ghost of a chance. The animals were only playing with him and he knew it. After an hour he dragged himself into my tent absolutely exhausted and flopped down. He simply radiated disgust and it was perfectly evident what he thought about all antelope.

But he got his own back one day. A herd of gazelle came within a few hundred yards of camp. Jumping into a car with Mac Young driving, we had a shot in less than a minute. I wounded two bucks but they could still run at a very respectable speed. Wolf had followed the car and had almost caught up with us when I fired. Without a pause he dashed by, running for the nearest gazelle. Mac and I jumped in the car to watch the finish. It was a beautiful race. Inch by inch Wolf hauled up on the buck. For a hundred yards he ran almost at its heels; then with a terrific spring he hurled himself on to the quarters of the gazelle. They rolled over together but Wolf had a hold on the flank and his weight held the antelope down. Suddenly shifting to the throat he tore open the jugular vein. When we had the antelope in camp Wolf lay down beside it and for hours would let none of us come near. It was his buck killed by himself alone, he thought. It represented final victory over the animals which had so often made a monkey of him before all the camp.

In spite of this, Wolf assumed responsibility for the safety of a baby gazelle which we had as a pet. One of the Mongols caught the little thing when it was only a few hours old and brought it to camp in the sleeve of his coat. Wolf knew instantly that that gazelle was not to be killed. It was so tiny and helpless that it had to be protected, and if our two ravens attempted to annoy it as they did everything else, Wolf would drive them off like a tiger.

Lieut. "Bill" Wyman, topographer, reared the baby on dried milk. It was a delicate job and only Bill's devoted care saved its life. Every two hours it had to be fed with warm milk, and Bill trained himself to wake at feeding time all through the night. He would light the candle, warm the milk, push the nipple into the gazelle's mouth and sleep for another two hours. Finally we bought a goat from the Mongols. They kept her kid but it was some days before she would let the gazelle nurse. Also the baby wanted none of the goat. At last they adopted each other and became devoted companions. But the goat's legs were so short that the antelope had difficulty in nursing. Bill Wyman solved the problem by building a platform of stones for the goat's fore and hind feet. It was screamingly funny to see the goat solemnly hoisted to its pedestal and stand chewing its cud while the gazelle ran underneath between the stones and greedily devoured its breakfast.

Wolf met his match one day in the shape of a hedgehog. As you know, these little fellows are about half the size of a hare and are covered with spines as sharp as needles. When they roll themselves into a tight ball they are absolutely impervious to attack. Wolf's first hedgehog drove him nearly mad. He couldn't figure it out at all. It smelled like an animal but it didn't feel like any animal that he had ever touched. Biting only got his mouth full of quills. Rolling it over with his paws pricked his feet. Barking didn't do any good. When it uncurled and started to run, at his first touch it rolled up again. For two days he puzzled over the enigma, assisted by a Mongol puppy and a pet crow. At last he definitely gave up. But he had lost so much "face," as the Chinese say, over the performance that he decided to ignore completely the existence of the darned thing. We kept it as a pet for months but never would Wolf so much as glance in its direction even though it actually ran over his feet. He would quickly move away apparently absorbed in some object in the far-distant horizon.

Another of the camp pets used to annoy Wolf a good deal. This was a black vulture. It is one of the largest birds of the world and has a wing spread of ten feet. Ours was named "Connie." She had been reared from a fledgling and of course was perfectly tame. Wolf had a particular place in the back of the tent, close against the cloth, where he loved to sleep. Connie liked the same spot. When she found Wolf in possession Connie would go to the outside and jump up and down on the cloth directly over the dog. Wolf would wake with a start and dash out only to find Connie hurrying to get inside before he returned. This is really true. I have the evidence of all the other men on the Expedition to prove it. You have to be particularly careful about your stories when there are fifteen other men who can check you up.

The bird had really amazing intelligence. We used to give her water from a gasoline tin which always stood at the entrance to my tent. Connie would stretch up her neck, open her great beak and take the water in gulps as one of us poured it for her. One day I was sitting in the rear of the tent, writing. Connie came up to the can and waited expectantly. I paid no attention and soon she hopped up to my chair, tugged at my coat and then went back to the can. I gave her water and every one was happy.

Vultures, you may not know, are among the best of all birds for pets. Because they are carrion feeders one thinks of them as being horribly dirty. But that is only because they have not had the advantages of a proper education. Environment can do anything. It can even turn a filthy vulture into a most clean and respectable bird. Connie used to take a bath two or three times a day if we were near a stream or salt lake and spend hours preening her feathers and drying out in the sun with wings half spread. I think she was the cleanest bird I have ever seen. We could not get her to touch carrion even when half starved. She had always been fed fresh antelope meat and nothing else would do. Once when we were in a country where there were no gazelle, she went without food for three days rather than eat some bad smelling scraps which N. C. Nelson, the archæologist, found for her. She was persuaded to swallow a bit of antelope liver but obviously hated it. I brought her to San Francisco on the U.S.S. *President Taft* and thence by train to New York. Of course she was a curiosity and always had an audience on the after deck. But she was uniformly good-tempered and it was touching to see the joy with which she welcomed me when I came for a visit. She would poke her great head under my coat and rub up and down nestling it under my arm.

I did the steamship company a rather low trick when purchasing her ticket. I said I wanted a ticket for a bird. The agent looked up the rates and found that the only birds listed were canaries and the passage cost five dollars.

"I suppose it *is* a canary, isn't it?" he asked.

"Well, no," I admitted. "It isn't a canary, exactly. Do I look like a man who has canaries?"

He laughed while making out the ticket and forgot to ask what kind of a bird it was. So Connie traveled on a canary's ticket and ate two pounds of meat a day. But the steamship company got more than enough publicity to pay her food bill.

When we camped in the desert three hundred miles from Kalgan, the Danish Minister to China, Hon. Henrich von Hoffman, stopped at camp for a visit while *en route* to Urga. He arrived late in the afternoon and we gave him a tent close to the edge of the escarpment. In the morning he opened his eyes to see a huge black vulture sitting on the foot of his army cot, solemnly gazing at him. He couldn't believe his eyes and finally yelled to me to ask if I saw what he saw.

I might go on *ad infinitum* with stories of Connie's amusing exploits, but Wolf is really the subject of this narrative and there is a good deal more to tell about him. He didn't like Connie but he tolerated her. Except when she drove him out from his place in my tent, he assumed a lordly indifference and disdained to even notice her. A Mongol puppy which we had, loved to play with the vulture but not Wolf. He was the most dignified and the most jealous police dog I have ever known. He simply would not tolerate another dog in the Peking house, no matter how small it was. In camp it took weeks to make him accept the Mongol pup. We acquired the puppy in rather an interesting way.

The Expedition stopped for water at a well on a caravan trail. Not far away lay the body of a dead horse. The birds had emptied the carcass and pulled off most of the flesh. Inside the thorax a little black pup, only a few weeks old, was living. Evidently it had been lost from a Mongol caravan. The puppy had made himself a comfortable bed inside his strange house and kept fairly well fed by eating the scraps of flesh that still adhered to the skeleton. Dr. Granger's car was the last to leave and as he was assisting the topographers, he stopped half a mile from the well. Thus he spied the puppy legging it after his car as fast as it could run. It was just about all in when it reached him and of course he couldn't leave it to starve on the desert. The poor little thing smelled horribly but Granger gave it a bath that night and it immediately adopted him.

Wolf resented its presence bitterly. Of course it was beneath his dignity to injure so small a dog but if it came near him he snarled and showed his teeth. As the puppy grew older it tried its best to play with Wolf. Its efforts were pitiful for it evidently looked at the big police dog with a sort of hero worship. Wolf resisted all advances for at least two months. Finally his resentment cooled and he would deign to play for a few moments in rather a shamefaced way. But he never became fond of his little Mongol companion or really relaxed his dignity in its presence.

Wolf had one bad habit that nearly cost him his life. He was a sheep killer. I say "was" but perhaps it would be better to say "is." It is almost an axiom that sheep killing can't be cured in a dog. I don't think it can, permanently, but I did cure Wolf temporarily. The method, however, was pretty severe.

The disease developed in Wolf rather gradually. The first year he was with us nothing happened until the middle of the summer. Then Walter Granger came into camp one night and reported that he had found two freshly killed sheep in a ravine and had seen Wolf not far away. Apparently the two sheep had strayed from the flock, Wolf had come upon them suddenly and was overcome by temptation. This first experience of such a diverting pastime gave him the germs of a disease which raged in his blood like a fever.

A few days later George Olsen actually saw him kill a sheep. The Mongol owner came into camp highly incensed but was pacified by receiving more money than the animal was worth. We ate the sheep. It became a common occurrence and the men complained of having too much mutton. But I will say that Wolf was a good judge of sheep so far as their edible qualities were concerned. He always picked tender ones. Finally we had to tie him up until we left that place.

At the next camp our caravan joined us. The loads were no sooner off and the camels peacefully grazing out on the plain than Wolf began to do his stuff. Selecting a camel right in the middle of the herd, he drove it out, biting and barking at its heels. When he had frightened the poor beast half to death and it was running across the desert as though the devil were after it, Wolf came back for another. In half an hour, in spite of all the Mongols could do, our camel herd was fleeing in every direction. We couldn't stop Wolf. He was deaf to all commands and no one could catch him. He was possessed of but one idea—to scatter that herd. Some of the camels were so frightened and ran so far that our Mongols did not find them for three days. Of course Wolf was tied up after that performance. But he seemed so penitent and pitiful that I let him free after a week's punishment.

Our next camp was on the edge of an oasis near a Mongol village. The first evening Wolf distinguished himself by tearing the throat of a heifer and scattering the herd of cattle to the four winds of heaven. I gave him a severe beating which certainly hurt me more than it did him and tied him up for another week. The climax came some time later when he killed two sheep half a mile from the tents. A big Mongol rode into camp absolutely furious. Before he calmed down I thought blood would be spilled, the worst of it being that he was perfectly right. We didn't have the ghost of a defense. He said that if we did not keep that infernal foreign dog tied up we'd have to leave right then or he and his friends would drive us out. It was a rich fossil field and we couldn't go. We tried another method that I hoped would work because of Wolf's great dignity. I tied one of the dead sheep to his collar by a short rope. Wherever Wolf went he had to drag that sheep. It was to be a constant reminder of his sins. Never did I see a more dejected or humiliated dog. He wouldn't look at any one; he was crushed to earth. After two days of living beside the dead sheep his condition was pitiful. But I thought for good measure that he had better have another twenty-four hours. Then we let him loose and I watched him carefully.

There were half a dozen sheep near camp. For five minutes Wolf ran from one tent to another visiting every member of the Expedition just as though he had returned from a trip abroad and was infinitely glad to be home. He seemed to want to dispel the idea that he had ever been so disgraced as to be tied for three days to a dead sheep. I had decided that either Wolf had to be cured or he must be killed. We couldn't stay in Mongolia and have the natives turned against us by a sheep-killing dog. Much as I loved Wolf, the Expedition was more important than his life. As a last resort I determined to give him a dose of fine shot. For an hour after his release Wolf was able to resist the temptation of those grazing sheep. Then, suddenly, he cracked. I saw him gaze fixedly at the animals and begin to tremble. Saliva dripped out of his mouth and like a shot he was off. Shouts were useless. He was in the grip of an uncontrollable disease.

I grabbed my shotgun and just as he sank his teeth in the throat of a young ewe, I fired a charge of No. 6 shot at his hind quarters. The impact knocked him off his feet, but he certainly was a sportsman. He never even yelped but limped off silently behind the camp. He circled about and tried to crawl into my tent. I hardened my heart and drove him away. He went to George Olsen's only to be refused asylum. Out on the desert he crawled and lay down a hundred yards from camp. I was almost weeping but he had to be cured or killed. After an hour I asked Dr. Loucks, the surgeon, to go out and examine him. He reported that Wolf's rear end was pretty well filled with shot but that no bones were broken and that he would be all right in a few days.

Next morning he was a sorry spectacle, so stiff and sore that he could barely walk. He got no sympathy or petting but by the end of the week he was pretty well recovered. It cured him temporarily. We had no more trouble that season. Whenever sheep were near camp Wolf would deliberately lie down in the rear of my tent, away from temptation. If I saw the suspicious trembling and drooling, a sharp word sent him into the tent. But the next season the old disease was back upon him, although not so strongly. He could resist it if one of us were near but he did kill two or three sheep before the summer ended. The treatment had one permanent effect, however—it made him deathly afraid of guns. Even in Peking when I was practicing with my revolver in the mornings Wolf would crawl under my bed until the shooting ended. He feared a gun so badly that it was next to impossible to get photographs of him. The moment a camera or a stick was pointed in his direction he ran like a stag.

The poor dog had another experience which he never forgot. The surgeon, Dr. Loucks, wanted to inoculate him against rabies before we went to Mongolia. He had to be muzzled and trussed like a pig for market and I never have seen such abject terror exhibited by any animal although he was not hurt in the slightest. In some way he associated it with the shooting in Mongolia. Like many surgeons, Dr. Loucks had a faint odor of iodoform about his clothes. From that moment on the smell of iodoform or the sight of an instrument case would send

Wolf into the gardens, up the stone rockwork and onto the roofs. He used to prowl for half a mile over the roofs of the neighboring houses, hunting cats and greeting the dogs in the various compounds; but he never descended except in our own yard.

He was ill one day and I telephoned Dr. Loucks to look him over. Wolf was lying on the floor in my office. The instant he heard me mention his name, he sat up listening intently. He would not go back to sleep and half an hour later when the doctor arrived, he took one look and dashed for his safe haven on the roofs. The Chinese servants all firmly believed that he could understand English and used to warn me not to talk about him over the telephone if he were present.

When a visitor came to the house, Wolf greeted him with delighted barks. He would sit up and offer his paw and every one instantly succumbed to his charm. But Wolf knew in a few minutes whether or not they were just being polite or whether they really liked dogs. I remember that Noel Coward, the actor-playwright, captured his affections at once because Noel adores dogs. Before I knew it they were rolling over and over on the floor having the most wonderful romp. While Noel was in the house Wolf would not let him out of his sight.

When the Expedition was in the Gobi, Wolf was a busy dog. Everything that went on about the camp was his concern. At night particularly did he feel his responsibility. As long as candles were burning in the tents, he would drop in every half hour for inspection. When the men crawled into their sleeping bags and extinguished the lights, Wolf would make a last visit, nuzzle each one to be sure that he was all right and then cross that tent off his list. After all the lights were out Wolf would lie down about ten feet in front of my tent and doze for a few moments. But always with one eye open. Once or twice during the night he inspected the camels, although he seemed to feel that they were not really part of his job.

One night a party of six bandits rode into camp. They made a grave mistake for they had seen only two tents. Wolf had the whole camp roused in three minutes and the bandits got the surprise of their lives when they saw lights flash on in fifteen tents. We took their rifles and let them go, much to Wolf's disgust, for he was aching to sink his teeth into various soft parts of their anatomy. Of course the bandits protested that they were soldiers. Probably it was quite true. In China virtually all soldiers are bandits. Receiving no pay and not too much food becomes boring after awhile and whenever a chance to rob presents itself, the soldiers change their status abruptly. Sometimes it is only temporary but often they take to the open road for good.

It was a sad time for both Wolf and me when I closed the Central Asiatic Expedition's headquarters in Peking. It meant breaking up the home in which we had lived for twelve years. And it meant saying good-by to each other for a year or perhaps forever. With the initial packing Wolf's excitement rose to fever heat. He thought we were getting ready for Mongolia. Then slowly he realized that something was wrong. This wasn't the usual kind of an expedition. There were no motors chugging in the great front courtyard and the rooms were being dismantled.

Sadness crept over him and he drooped like a half dead flower. The last week when an auction sale was held in the house Wolf was really ill. For two days he touched no food. He drank water in great gulps but I could not tempt him to eat a mouthful. He lay on the floor in my bedroom, gazing sorrowfully at me with his big brown eyes. If I moved he was at my heels. He did not want to leave me for a second in what he knew perfectly well were our last hours together. The day before I left Peking I took him in my car to the house of one of my friends, H. R. Ekins, manager of the Peking branch of the United Press. Wolf ran about, inspecting the rooms but always coming back to gaze into my face. In the hall when I put my arms about his neck to say good-by his head was wet with the tears which I could not control. I never saw him again.

You may wonder why I did not bring him to New York. It would have been cruel to him. All his life he had roamed the plains of Mongolia or had a great Chinese compound in which to run. For him, in New York, life would have meant my penthouse roof; lonely days when I left in the morning and only a few brief hours together while I was dressing for dinner. It would have broken his heart—he would have died in six months. Far better that he stayed in China in the surroundings where he was born, with a new master who is already his devoted slave. If ever I return, Wolf of Mongolia will be waiting and I know that I shall not be forgotten.

CHAPTER VI

Superstition

RELIGIOUS beliefs are certainly the most fruitful source of potential trouble when an explorer is dealing with primitive natives. Ideas which to us are laughable in the extreme may be of the utmost seriousness to a native. Almost all primitive religions are based largely on superstition; in their operation there is comparatively little deep religious feeling such as we understand it. Natives believe that if such and such a thing is done, the result will be immediate; it may come in after life also, but they are morally certain that tomorrow or next day unfortunate things will begin to happen. They think of their God much as a human being; if they make Him mad He'll get back at them pronto. I could give a score of instances where explorers have met disaster for their expeditions or death because of such ignorance. Therefore it behooves an explorer to learn about the religious beliefs of any natives he is to come in contact with and to watch his step if he expects to avoid trouble.

Like most lessons I had to learn mine by experience and I confess that I was rather stupid about it. Some were amusing and some might have been serious. One of my first experiences was on the edge of Tibet. I wanted photos of the people but they ran at the sight of a camera. I decided to get the photos anyway. We were camped near a trail where Tibetan caravans were continually passing on their way to the tea regions in the south. I hid behind a bush and when a particularly picturesque native came along I leaped out with camera poised. With a yell they would dash behind a mule. One chap trudging along beside a woman grabbed her around the waist and held her in front of him while he backed off. I didn't learn until afterward that these particular people believed that I would take away their spirits in the black box; that lacking the intangible something that made their bodily wheels go around, they would soon shrivel up and die. Moreover, that there would be no Happy Hunting Ground for them in the future life.

After a couple of days of this stalking of Tibetans, my men learned that the natives were about to descend upon us with murderous intent and that the quicker we cleared out the better. It was only then that I realized that what was rather silly or amusing to me was far from either silly or amusing to them and that through ignorance I had committed a grave social error.

So again on my first trip to Mongolia I let myself in for an experience which might have been serious had it not happened in the house of a Mongol friend. The Mongols believe that the instant life departs evil spirits take up their abode in the corpse. If any one touches a dead human body, or even the bones or skull, he draws the evil spirits into himself. I did not know this. When I first went to Urga I lived in the house of a Mongol Duke, Lobitsan. He was a splendid fellow, a keen sportsman, full of humor and much liked by foreigners.

I was shooting ducks one day near the Tola River

and found a human skull. It was clean and white. I knew that the Department of Anthropology of The American Museum of Natural History needed Asiatic skulls so I put it in a game pocket and returned to the house, very pleased with myself. Lobitsan was in the room when I produced the skull. He leaped to his feet, furiously angry, shouting at me in Chinese and Mongol and ran into the courtyard.

"Take your skull and get out of my house," he yelled. "You can't stay here with that. Get out! Get out!"

I grabbed it and dashed for the compound gate, for Lobitsan was beside himself with rage.

One of my friends who had lived in Mongolia for years fortunately was there. I was completely bewildered but he explained what it was all about. He said that I must take the skull back to the exact spot where I had found it; in the meantime he would see what he could do.

Two hours later when I returned there was the most infernal din in the house. Cymbals clashing, drums banging, and through it all a continuous drone of lama voices reciting prayers. My friend met me outside. He said that the Duke had rushed to the nearest temple. The high priest had sent a riot squad of lamas and "sicked" them on the evil spirits which I had let loose in the house. This prompt action probably would save the place; otherwise it would have to be burned.

They were like a force of American insect exterminators. For two days and nights the incantations went on continuously. Then the high priest announced to Lobitsan that he had the spirits pretty well under control and that he felt sure none were lurking even under the floor boards. A thousand dollars would be his fee. Also I had learned my lesson about Mongol superstitions.

Until Lamaism was introduced into Mongolia from Tibet and became the official religion about the time of Kublai Khan's death in 1294, the natives were Shamanists, worshiping the gods of the hills, rivers, rocks and plains. Much of this nature worship still remains and of course all their religion is based on superstition. It is the great weapon by which the lama priests keep their people within their power. One time when I was camped in the forests of northern Mongolia not far from a Mongol village, a wandering lama dropped in for a few days' visit. A baby had a bad case of eczema. I cured it. A man had a dislocated shoulder and I put that back in place. Hardly a day passed without some native coming to me with an ill of some kind. I made plenty of cures but the lama priest got all the fees because in every case he told the patient that the cure really had been effected by his prayers. They admitted to me, in private, that they really believed it was my foreign medicine that had done the trick, but that they didn't dare to anger the lama or he would bring on them dire curses—that their sheep and horses would die and they themselves would perish of awful diseases.

Of course any one who has read Kipling's delightful story "Kim" has rather a sentimental feeling about lamas. I suppose that there are some priests like his, but those whom I have seen are as a whole the most ignorant, debauched and completely undesirable parasites that I have ever encountered. The lamas, and the Chinese too, think that they can fool their gods. In my penthouse in New

York I have three lovely bronze lanterns which I got from a Tibetan temple. The priest wasn't very keen to sell them to me but finally he succumbed to the bait of empty bottles. Before he took them from their places, however, he plastered over the eyes of the Buddha with a handful of mud. I said, "What are you doing that for?"

"So the God can't see me taking them out."

"But," said I, "he could hear us talking about it."

"Oh, no," he answered, "we are speaking Chinese and he can only understand Tibetan!"

As a matter of fact, that isn't confined only to ignorant natives. A friend of mine who was professor of literature in a Christian university in Peking once asked his class of Chinese students why when they sang religious songs they always used English words. One boy answered, "because Jesus Christ was a foreigner and he wouldn't understand Chinese!"

Usually the reaction of natives to something they can't understand is fear. Often fear can be used to great advantage by an explorer. But he must realize also that he can't scare the people of the country half to death and expect the friendly coöperation which may be necessary for the success of his expedition. There are times when it is impossible to avoid frightening natives or when it is completely unexpected.

I remember that some years ago I was crossing the Gobi Desert from Kalgan in north China to Urga, the capital of Mongolia. The Hutukhtu, or Living Buddha, wanted an electric light plant and one of my friends had sold him a Delco. We made camp about a mile from a Mongol village of a dozen *yurts*. Ted McCallie suggested that we rig an arc light on a pole and give the Mongols a thrill. It seemed like a sound idea. The arc light was duly rigged and blazed out like a beacon on the seacoast. We sat ourselves down to wait for the influx of Mongols whom we had aimed to impress. But no one came. Next morning in place of the Mongol village there was a tenantless plain. Not a sign of a camel or a sheep or a *yurt*. They had all moved out during the night, frightened absolutely pink. Word of the strange ball of fire that had suddenly appeared in the desert traveled even faster than we did and we heard of it from natives hundreds of miles away.

I think that the most amusing experience I ever had with Mongols was when the Central Asiatic Expedition was encamped in the Gobi Desert not far from a large temple. Few of the natives ever had seen a white man. Anyway when our motor cars roared up they were frightened nearly to death. All those who had ponies leaped on their backs and fled to the hills. Eventually they drifted back and when they discovered that we were friendly and harmless they crowded about in silent, staring groups.

I started to shave but the tent door was so packed with Mongols that I finally decided to perform the operation in public and moved out into the open. In the first place a Mongol has very little beard and to see me lathering my face and scraping off the hair with a safety razor, which didn't look like a knife, was something to write home about. They were so pleased that I called one of our men who had two sets of false teeth, to give them a real show. He rolled up his sleeves, spread out his hands in the most approved manner, and pointed to his teeth. Then with a yell he yanked out both sets and held them up for inspection. The Mongols nearly fainted. He clapped them in again and grinned. We told the natives that any of us could do the same thing. Who would they like to see perform next? Meanwhile I shoved another man, Albert Johnson, up into the front row. Albert also rejoiced in the possession of false teeth. A big Lama pointed to Albert and he pulled out first his upper and then his lower set. That got a big hand. But the climax came when we produced our archæologist who has a glass eye. Very impressively I explained to the breathless audience that we weren't limited to teeth but could even take out our eyes as well. Would they like to see? They would. Out came the eye; it was exhibited and was then popped in again.

By that time the Mongols were ready to believe anything. They asked if we could take off our legs and arms as well. I said of course we could but that we were getting a bit tired and would have to call it a day. Our reputation was made, however, and it traveled far. The next year when we were hundreds of miles away some of our Expedition were camped in a gloomy canyon at the base of the Altai Mountains. A family of Mongols stopped for a day on their trek to other grazing grounds. They hailed our men with enthusiasm for they said every family in the Gobi Desert had heard of the wonderful way in which we could take off various parts of our bodies and put them on again. Could they have a demonstration? There was nothing doing, however, for the men with the false teeth and the glass eye weren't with us that season.

Natives of different countries by no means act alike. Martin Johnson tells me that the natives of Africa are not particularly interested in aëroplanes nor are they really frightened when they see one. They simply accept it stoically as a part of the white man's way of doing things. If he wants to fly in the air it's all right with them so long as they don't have to do it themselves. They don't like that a bit.

In Mongolia, however, the natives certainly did not take our motor cars stoically. They were frightened nearly to death. Whenever we neared a Mongol village we would see ponies dashing away, their riders beating them in a frenzy of fear. And those same riders were always men. Only women and children remained in the *yurts* when we arrived. As Shackelford, our photographer, said, the Mongol's motto is "Save the men, to hell with the women and children." In one village a woman who was kneeling on the ground holding a huge dog, became violently nauseated from sheer fright when she saw Walter Granger. I never ceased teasing Walter about that. Another time as we dashed up to a *yurt* two women met us at the door, spreading out a felt mat and offering us a pitcher of milk. The younger was the only really beautiful woman I saw in all Mongolia. When they learned that we were friendly they cried on each other's shoulders in relief. They confessed that they had expected to be killed at once.

The men who had fled would drift slowly back, when they saw that murder wasn't to be done, wearing rather sheepish expressions. And yet the Mongol isn't a coward. He is "a first class fighting man" and when he starts anything he is pretty likely to see it through to a finish. But a motor car is something he never dreamed of and can't understand. Coming with such a rush and roar I don't wonder that it is terrifying.

The Chinese call them *chi chur,* "wind carts," and the Mongols always supposed that it was the whirling fan sucking in air which made the car move. Our favorite sport was to wait until a group was closely packed about the front of the car watching the fan and then to sound the horn. The results were invariably satisfactory—natives tumbling over on their backs and flying in all directions. But the Mongols enjoy a joke, even upon themselves, and it all contributed to good feeling toward the members of the Expedition.

The Mongol reaction to our Victrola was interesting. A musical record got no results because it was simply a noise—our music doesn't sound like music to them. But when we put on a talking record things happened. The sound of a human voice coming out of the box started them off in a panic at first. When they found it was harmless they would sit for hours listening to it.

It is always interesting when an explorer meets people who have never seen a white man. Of course one can see how they act but what they really think is difficult to know. I always expected that they would be tremendously impressed and mystified by cameras, watches, compasses, and gadgets of that sort. But it doesn't work that way really. Among the first natives I met who had never seen a white man were the wild Lolos of Yunnan Province 'way up on the frontier of eastern Tibet. We dropped into their village one morning literally out of the clouds. Down from a sixteen-thousand-foot mountain pass we straggled after a freezing night on the very summit.

The Lolos lived there in a secluded valley. They were shy but not frightened, for a Lolo doesn't scare easily. A fight is the breath of life to them. They were curious about us and all our things, of course. Watches were completely beyond their understanding and they weren't interested. Also cameras, at first. I took some photos of them but the subjects didn't recognize themselves for they had never seen their own faces. It was only when I pointed to various articles of dress, which they were wearing and they saw them in the photographs that a dim understanding began to percolate into their primitive brains.

Mirrors made a great hit and bottles were almost priceless. One man offered to trade his number two wife for an empty bottle. I said that I would only be interested in number one who was a rather attractive girl, but he explained that she was a very good worker and moreover was expecting a baby soon so he couldn't let her go because I'd be getting double value.

What really gave them a kick were things which they knew something about, such as rifles. They had the most primitive flintlock guns which would only kill at about sixty yards. One day I shot a sheep with my Mannlicher 6.5 mm. high power rifle at three hundred yards. I showed them the tiny cartridge first and then shot the sheep. When they saw the animal drop and looked at the great hole in the side torn by the mushroom bullet, they could hardly believe their eyes.

My high leather boots they loved and a rubber rain coat impressed them beyond words. Actually to pour water on a cloth without its being wet was magic. Field glasses, of course, are a never-ending source of wonder to all natives everywhere. The Chinese call them the *chen li yen,* "thousand-mile eyes."

It all boils down to the very simple and natural fact that the only things by which they are impressed are those which they can understand, and which have to do with their own lives. Incomprehensible things are either frightening or are as uninteresting to them as the Einstein Theory of Relativity is to the average person.

But my actual body was always a source of wonder. I am blond and have blue eyes. Most Orientals have brown eyes. When the Lolos discovered that I was real and could talk and eat and breathe and laugh just as they could, they gave themselves up to finding out just what sort of a being I really was. With eyes as light-colored as mine, they couldn't believe that I could see properly. It was a tremendous surprise when they discovered that I wasn't blind. Of course my face was so tanned by wind and sun that it was almost as dark as theirs. But when I showed them the white skin of my body and legs they simply gasped. And what a hit I made with the women!

As a matter of fact natives themselves have some things that to us seem mysterious. We hear a lot about the "grapevine telegraph"—the strange methods by which natives communicate with each other over long distances. Even in a prison I'm told information seems to percolate through the walls—and certainly I know it does in a natural history museum. Just try to keep anything secret in The American Museum of Natural History! I'm the Director and I know that it can't be done. How it works in a prison I don't know and I hope that I won't find out from personal experience. I *may* discover the secret in my own institution but I doubt it. But I can tell you how it works in the Gobi Desert, and there isn't anything occult about it. No "talking drums" as in Africa—just horseflesh and curiosity.

Wells are the great meeting places in the desert. Here natives gather to water their sheep and goats and horses and camels. It takes a long time, for the water has to be pulled up hand over hand in a single skin bucket. Perhaps half a dozen families may gather at the well. If a Mongol happens to be passing four or five miles away he rides over to the well to gossip and hear the news. Coming from all points of the compass anything that has happened is carried in as many directions when the natives leave. If it is something really unusual one of the men is sure to ride to the nearest *yurt* to spread the news. That may be thirty or forty miles away but the Mongols have little to do and a choice bit of gossip is more than incentive enough. In an extraordinarily short time anything that is happening on the desert is carried hundreds of miles.

Once when I was three hundred miles from the main trail between Kalgan and Urga I profited by it myself. I heard of a motor car that had been seen going fast across the desert carrying a man tied hand and foot in the back seat. This was just two days after the car had passed. The Mongols didn't know what it was all about but I did. It happened to be the end of a drama which had begun while I was in Urga.

An American, Williams by name, an employee of a trading firm, had arrived in Urga with his wife and child. In the city at the time was an Austrian named Kunhardt who was a typical bad man, suspected by the Mongols and their Russian advisors of half a dozen murders. Kunhardt was a prisoner at large in the city while the Mongols were collecting evidence against him. One night he got Williams drunk and induced him to sign a paper stating that he, Williams, pledged his own life and that of his wife and child, for Kunhardt's return if the latter were allowed to go to Kalgan. Kunhardt left. Poor Williams was in a bad mess. The Mongols would not allow him to leave Urga to bring back Kunhardt who had no intention of returning to eventual execution. Williams was advised to escape from Urga and get Kunhardt, which he did with the aid of a German chauffeur.

In Kalgan the two men kidnaped Kunhardt, tied him securely and started across the desert for Urga. During the ride Kunhardt complained of the tightness of his bonds and the ropes were loosened just enough to enable him to slip his hands free. Williams and the German were in the front seat. Kunhardt made a sudden leap, snatching at a knife in Williams' belt but he saw him coming in the windshield just in time. Kunhardt was bound tighter than ever and the day after arriving in Urga he died of gangrene in his arms and legs.

Williams was received with open arms by the Mongols. He had returned of his own free will bringing Kunhardt with him. Kunhardt was conveniently dead and that saved the authorities the trouble of shooting him. Every one was happy. I was particularly so, for Kunhardt had murdered a Danish friend of mine by the name of Olufsen and I had sworn to shoot him on sight.

CHAPTER VII

J. McKenzie Young—Explorer

YEARS ago Admiral Peary said to me, "The three most important qualities which an explorer must have are loyalty, unselfishness and dependability."

As leader of many expeditions, I have found that Peary was right. I have selected my men always with those characteristics in view. Once I refused against continued pressure to take a man with a brilliant scientific record because he was selfish. He was one of the two best men in the world for his job except in that one particular.

Another man whom I should have liked to take was ruled out because I knew that he never would obey orders if he did not approve of them. The leader of an expedition is responsible not only for the success of the work but for the lives and safety of his whole staff. Presumably he knows more about the job than any one else. His men never should go unless they believe him fitted to be a leader by experience and temperament. Once in the field, his authority must be absolute. I don't mean that he should not consult with his men. He should hear all views and consider them carefully. But the final decision must be his and the men must recognize the fact.

I have had some wonderful men with me in the desert. Men who would die for me if necessary; who trusted me so implicitly that no matter what I asked they were willing to do it, believing that behind it all there was a good and sufficient reason.

One of the best of these was McKenzie Young, soldier, adventurer and my friend. This is a sketch of his life and his tragic death. Kipling might have written his poem "The Men That Don't Fit In" for Mac. Restless, a wanderer, searching for adventure and always finding it.

Once I said to him, "Mac, why don't you settle down? A rolling stone gathers no moss, you know."

"Who wants moss, anyway," said he.

And that was true. Moss and Mac Young never could go together. The story I am telling you now I heard from Mac in bits as we rode together across the Gobi Desert or slept at night on some lonely peak of the Altai Mountains.

Mac's father is a charming white-haired Scotch clergyman, and his mother is everything a mother should be. Mac's early days were spent in Pittsburgh, but he was in school in Canada when the world went mad in 1914. One by one his friends enlisted. Mac couldn't stay away. He obtained a commission in the Canadian forces. But he wasn't sent across to France where he wanted to go. A general who was in charge of recruiting took him as his aide. Mac was handsome as an Adonis, and charming in personality. For months he fretted at his job. Finally he could stand it no longer.

"General, when do I go across?"

"You don't go. You are too valuable to me here. I need you," was the answer.

"I didn't enlist to fight the war in Canada," Mac told me. "I damned well wasn't going to be stuck as aide to a superanimated old general. So I deserted and joined an outfit then on the way to France. In England I arrived under arrest."

Court-martialed, the sentence of the presiding judge was: "You shall be sent to the front at once. I admire your spirit, but deplore your judgment. You never again can hold the King's commission."

To France Mac went. His outfit was sent immediately to the front. Mac had a day and a half in the trenches. Attacking at dawn just as they went over the top, he stopped a piece of shrapnel in the leg. Back to the hospital base; then to England to recover.

Again in France and again wounded almost at once. England a second time, spending long weeks in a hospital. Back to France, this time as one of a battery of six-inch howitzers. During a year of fighting the big guns, Mac had the awful experience of being buried in a dug-out by the explosion of a thirteen-inch shell. He got out just alive, the only one of his battery to survive.

"That was worse than being wounded," he said. "There was something terrifying about it. I felt like a rat in a hole. I couldn't stand a thing like that again."

When the United States joined the Allies in 1917 Mac wanted to transfer to the flying corps of the American army. But the Canadians didn't want to lose him. He had risen to the rank of top sergeant and that was as high as he could go. But he persisted and finally got his transfer. He was in air training when the war ended.

Back in America, Mac went into a bank. A bank of all places, for a seeker of adventure! Of course he couldn't stick it very long. So he wandered northward to Canada, lured by the legends of the Northwest Mounted Police. But it wasn't what he hoped it would be. He was too valuable at headquarters. For several years he was a "mountie" but did not reënlist.

There at the edge of the great country stretching northward to the Arctic Ocean tales came down of trappers and fur traders in the Mackenzie River region. Mac and two of his buddies decided to seek their fortunes in the north. None of them knew much about trapping but they could learn. They did learn too and it was a bitter experience during a long winter. They got plenty of furs, but Indians stole a cache of food. They struggled to a trading post just on the verge of complete starvation. The trader did not run true to the traditions of the north. In return for enough food to take them out he made them give him their best furs. At last they arrived in Seattle and the remaining skins were sold. Eight hundred dollars apiece was the net profit.

Of course they had a night of celebration. Until I had been on a long cruise myself and away from civilization for many months I never could understand why sailors at the end of a voyage want to raise Cain and spend every cent of money they have worked so hard to make. I found out because that was just what I wanted to do. Every shop looked enticing; every girl was beautiful; all music was intoxicating. The contrast and the sudden change upsets one's sense of values. You are happy to be back and you have to show it or burst. It is a natural human outlet, just as a volcano explodes when too much steam has accumulated.

Mac awoke next morning in a hotel with all his

money gone. He never knew just what happened or how he got there but the fact remained that he did not have a nickel. It was one of Seattle's grayest days, than which nothing can be grayer. Mac was hungry, his head ached like the devil, his spirits were far, far below zero. He passed a Marine Corps recruiting station "Join the Marines and see the world."

"Well," thought Mac, "I can always go back to the army."

"What's your experience?" asked the recruiting officer.

"Four years of the war. Big guns. Six-inch howitzers."

"Sure, we want you."

"What post is farthest from the U. S. A.?" asked Mac.

"Peking, China, the Legation Guard."

"All right. I'll join if you'll send me there. But you've got to promise. China for me."

So as an enlisted man for three years Mac joined the U. S. Marine Corps. But he didn't get to China at once. A station on the Pacific coast, training men to handle artillery, was where he landed first. As usual, he was too valuable; they didn't want to let him go.

But Mac held them to their promise and one brilliant day in autumn he arrived in Peking. Colonel (later General) Hal Dunlap was in command. Hal was one of my most intimate friends. We shared a temple together, which rejoiced in the name of "The Temple of the High Spirited Insects." Colonel Dunlap soon discovered that Mac was an expert motor mechanic, promoted him to Corporal and put him in charge of all the Legation Guard automobiles and trucks. I saw Mac often at the Insects temple. I needed a man to take charge of our cars on the second expedition to the Gobi Desert. Colonel Dunlap suggested Mac as I hoped he would. We got him assigned on detached duty to the Expedition and thus began our friendship.

For years Mac and I worked together in the field, sharing the joys and disappointments, the hardships and the dangers of an explorer's life. In Peking he lived with me at the Expedition's headquarters, having a courtyard of his own with its surrounding apartments.

While the Expedition was in the main camp Mac and I used to take a car and explore the advance country, seeking new regions in which to work. Thousands of miles we traveled together alone. I remember one time we had to drive from the very center of the Gobi Desert to Urga four hundred miles away over unexplored country. It was the 24th of May when we started, warm and beautiful. That night a blizzard struck us. We pitched the tent and crawled into the fur sleeping bags. Our clothes were piled on the ground beside us. We were wet and cold and those warm dry bags felt like a bit of heaven.

When I opened my eyes in the morning I could see only a white blanket. My body felt as though some one were sitting on my chest. It was several minutes before I realized that I was completely buried in a snowdrift. Eventually I got my head out of the white mass. The tent was packed with snow in a long drift sloping up from door to back. Mac was nowhere in sight. Suddenly the drift burst open and Mac's head emerged.

How we did laugh and curse as we dug our cold wet clothes out of the snow and got them on our shivering bodies! It was hours before we could dig out the car and get the engine started. Then began a day of discouragement and heartbreaking toil. Taking a compass course for Urga we started across the desert, plowing through an expanse of unbroken white. I was driving when suddenly the car seemed to go from under me in a sickening drop. We had driven over the brink of a deep ravine packed bank full with snow. All day we worked, digging the car out only to drop into another ravine a few miles farther on. That trip to Urga was a nightmare. We arrived eventually, half frozen, utterly exhausted and nearly starved. Only Mac's superb strength and his indomitable will had brought us through.

Mac's loyalty to the Expedition and his devotion to me were his religion. I went to New York one winter, leaving Mac in charge at Peking. He cabled me that word had come down from the plateau where our caravan was wintering that a certain bandit chief had stolen all our camels. The brigand was an old friend of mine and I felt sure that he did not know that the animals belonged to me.

I cabled Mac, "Go up to Mongolia, find the bandit and tell him that he has stolen our camels. I'm sure he'll give them back."

Mac went with one companion. It was the usual cold of a Mongolian winter—forty below zero. He found the brigand and explained the situation.

"Of course I'll return the camels. I didn't know they belonged to An Lao-yeh."

And so the matters were arranged. Mac spent the night in the bandit's camp. But he wasn't very comfortable there. They were a hard-bitten crowd of cutthroats and he was anxious to get away. A hundred miles lay between him and Kalgan at the railroad. Next morning there were flurries of snow drifting out of a gray sky. It was plain enough that a blizzard was in the making. The brigands urged him not to start, but he drove away. Hour after hour the snow whirled in blinding gusts about his car. Just at dark, after twelve hours of bitter cold, he reached Chang-peh-hsien, a walled city ten miles from the Pass. At a Chinese inn where he got out of his furs beside a fire he realized that the fingers of both hands were white and numb. Then began a night of torture. In the morning the Chinese tried to dissuade him from attempting to drive down Wan Chuan Pass which drops three thousand feet to the lowlands and is dangerous enough under the best of conditions. But Mac was nearly mad with pain. He had rather die, he said, than stay there another night.

The Pass was a hell of ice and snow. How he got down with that car is still a mystery to me. Only Mac could have done it. Eight Chinese lost their lives that day when the motor they were in went over a cliff. At Kalgan, Mac waited a night and a day for a train, suffering the tortures of the damned. Then in an open steel freight car, packed with Chinese like sheep, he rode fourteen hours in below zero weather to Peking.

When he reached the city Mac was completely out of his mind. Instead of going to the hospital, he wandered into the Club, babbling of a cable to tell me that the camels were safe. At the hospital the doctors believed it necessary to amputate all the fingers of both hands. Gangrene and certain death would follow if the black stumps were left.

Mac said, "No, I'd rather die than go the rest of my life with only two thumbs. I do everything that makes living for me worth while with my hands."

Doctor Harold Loucks, the Expedition's surgeon, examined him carefully. His blood was pure, his body as hard and fit as a trained athlete.

"Mac, there is just a chance that I can save most of your fingers. It is one in a thousand. But it will mean weeks of pain."

"Let's go," said Mac with a grin.

He did not escape the pain. It went on for weary months. When I returned his hands were swathed in bandages. Night after night I would find him pacing the courtyard. Together we would walk in the moonlight until from sheer exhaustion he could sleep. But no one ever heard Mac complain.

"I'm better," was the invariable reply to his legion of sympathizers. All Peking paid homage to his splendid courage. Seven months later when the last operation had been performed, only the ends of four fingers were taken away, and he could use his hands almost as well as ever.

The next year we went to the Gobi. In the autumn Mac and I were coming down alone in two cars. We had been warned that the trail swarmed with bandits. When we passed the Mongol village belonging to our caravan men, one of the boys ran out to signal us. He said that the previous night thirty brigands had killed two Chinese and robbed their cars only ten miles south on the road. They might still be there; he did not know.

Mac and I went on with our rifles and revolvers ready for action. We weren't asking for trouble, but we certainly did not intend to be driven off the road by thirty Chinese bandits. The cars had been held up near a mud house which had long been a brigand rendezvous. When we arrived, all was quiet and the place seemed deserted. But that was the usual procedure and we expected a fusillade of shots any second. Nothing happened, however, and we reached Kalgan without difficulty. A week later Mac went back. He drove one car and Liu Hsi-ku, one of our Chinese, the other. I had a presentiment that something would happen and asked Mac to be particularly careful on the road. Two days of rain had made the trail like grease. He fought mud all the way where we had driven over a hard dry terrain.

On the second morning two Mongol children told him that bandits were robbing a caravan just ahead. It was the familiar place of the mud house. The ground was so soft that Mac could not leave the trail and circle over the surrounding hills. He either had to go on or turn back.

"To hell with them," he said to Liu. "We're not going back through all that mud. If they want a fight they can have it."

The mud house appeared in the valley, half obscured by a train of oxcarts. Several men were going through the loads. Mac drove down the hill followed by Liu. He had nearly passed the house when from behind a mud wall thirty yards away three Chinese opened fire with Luger pistols. Bullets sang all about him, but he wasn't touched. He slowed down, swung about in the seat, and took a snap shot at one fellow who was doing the most effective shooting. His bullet struck a small stone in the mud wall an inch from the man's head. Either the steel jacket or fragments of rock tore off half the bandit's face. He fell backward, but the other two kept on firing.

Mac dared not take his hands from the wheel, for the car was skidding dangerously. Holding his rifle in one hand like a pistol, he fired three shots. One of them nicked another bandit.

In the meantime a dozen brigands, standing near their horses on the other side of the road, began shooting with rifles and pistols. Some had mounted and were riding after him when Mac slowed down, took a good aim, and killed a horse. That ended the matter. Mac was bad medicine. The other bandits galloped away.

When the Expedition returned a month later, they learned that there had been eleven brigands in the mud house. They expected an easy time when the two cars approached and got the surprise of their lives. They told the road police, who most of the time are bandits themselves, that they didn't know it was the "American Men of the Dragon Bones" as we were called, or they would never have opened fire. From previous experience they knew that our men all could shoot much too straight.

Unselfishness was one of Mac's most outstanding virtues. Time after time in crossing the desert together, when water was short; when the sun had turned the sand into a glaring furnace; when our tongues were swollen and our mouths like cotton, I have had to watch Mac to make sure that he took his share. I have seen him put the canteen to his lips, make gurgling noises and wipe his lips.

"God, that was good. Have a real drink now, Roy."

"Mac, you infernal liar, you never took a swallow. You can't fool me. I was watching your throat."

"Well, what's the difference? You need it more than I do. Your tongue is so thick you can hardly talk. See, I'm much better."

It takes a *man* and deep affection to do that sort of thing. Did you ever suffer from thirst? I hope you never will. It's pretty awful. Starvation is nothing compared to lack of water.

I have slept with Mac on the summit of the Altai Mountains when the cold bit like a knife and waked to find myself wrapped in our single blanket while he lay shivering without a cover. Those are the things that one never can forget; the things that endear a man to his fellows as nothing else can.

Mac and I had all sorts of experiences together. His insatiable desire for adventure and his priceless sense of humor made us inseparable companions both in the field and out of it.

We were in Urga together in 1925 after the death of the Hutukhtu or Living Buddha of Mongolia. We learned that special ceremonies were to be conducted upon the installation of his embalmed body in a great temple. We wanted to go, but it wasn't allowed. No foreigner would ever be admitted to the sacred precincts of the Holy of Holies. A young Mongol friend of ours, Dalai, said that he would get us in if we dressed as natives. He was taking a big chance in doing it for certainly we all would have been killed if it had been discovered.

Mac and I were both so brown from the sun that we needed only a little stain to make us as dark as any Mongol. Dalai borrowed some clothes and we sallied forth. Several hundred Mongols were already in the temple when we arrived. To a priest at the door Dalai explained that Mac and I were Mongols from the Alashan Desert who had journeyed to Urga, like many other pilgrims for this ceremony; that we were both deaf and dumb, a not very uncommon thing in Mongolia. The priest welcomed us courteously and we passed inside. I was a little worried about my blue eyes—Mac's were gray—but the temple was lighted only by candles and in the half darkness they got by.

The danger made it a thrilling adventure although the ceremony itself was not very different from what I had seen in other temples. At the far end of the room, facing the entrance, sat my old friend the Hutukhtu on a golden throne. He looked natural enough to speak except that he was only half his normal size and was completely gilded with gold leaf. His features were perfect. It was the finest embalming that I have ever seen.

On the right side of the throne sat a high priest dressed in a gorgeous robe of gold thread with a yellow Roman helmet-like hat. In front were two double rows of seated lamas facing each other. Between the pauses of the high priest's prayers the voices of the seated lamas swelled into a barbaric chant broken by the clash of cymbals. The air was heavy with burning incense.

Mac and I kneeled with the others, touching our

foreheads to the floor between our outstretched palms. It was a devilish uncomfortable position and I kept peeking out at Dalai hoping that we could get up and get out. But Dalai was immovable. We stayed for half an hour doubled up like jackknives until I thought my neck would break. Then I felt a poke from Dalai's foot and the three of us slowly sat up. Boy, what a relief it was! Other Mongols were rising too and we joined a group which drifted out of the temple into the courtyard. Then each of us had to whirl a dozen prayer wheels, murmuring the Tibetan invocation *Om mani padne om,* "The Jewel in the Lotus." Thousands of Mongols were streaming into the temple as we left, for the ceremonies continued all day long.

Back at Dalai's house we washed the stain from our faces and hands and pulled on our own clothes. It had been an interesting adventure and we had been where no other foreigners had ever gone. Our lives would not have been worth a plugged nickel had we been recognized. Of course that was what made it fun.

Mac and I were invited to dinner a few days after the temple ceremony by the Prime Minister, a fine old Mongol whom I had known for several years. He lived in a large clean *yurt* pitched on the bank of the Tola River not far from the Living Buddha's palace. It was a typical Mongol dinner. In the center of the *yurt* was a huge iron bowl of boiled mutton. We were seated on the ground facing the door as guests of honor. According to custom the host reached into the pot and presented to me the skinned head of the sheep. I knew what was coming and mentally braced myself. The two great eyes stared up glassily. I had to dig one out with the point of my knife, pop it into my mouth and make appropriate sounds of gastronomic enjoyment. My stomach rebelled but I had to go through it or mortally offend our host. I got the infernal eye out at last and crammed it into my mouth; then passed the skinned head to Mac. He looked as though he was about to bolt for the door.

"Go to it, old timer. Dig into it," I murmured out of the corner of my mouth.

Mac dug and choked, but managed to keep it in. I watched my chance and when the others weren't looking transferred the wretched eyeball from mouth to pocket. Mac's followed the same course. Then came the next ordeal. The tail of the sheep, a mass of solid white fat, was presented to us. But that wasn't so bad. We only were required to slice off a chunk and pass it on to the next man. These preliminaries ended, we could dip into the pot ourselves and personally select the portion of the sheep's anatomy we wished to eat.

The drink was buttered tea, like that of Tibet. Boiled tea in which butter, none too fresh, had been churned. After dinner we drank *kumiss,* fermented mare's milk. Not too bad, *kumiss*. Slightly acid in taste, cool and rather refreshing, but packing an awful wallop if one drinks too much of it. On our way home Mac and I solemnly produced the sheep's eyes from our respective pockets and with the greatest satisfaction threw them as hard as we could against a stone slab. They spattered beautifully!

J. McKenzie Young—Explorer

Mac returned to New York from China in 1931. The work of the Expedition was temporarily suspended and I remained to close the headquarters. In mid-August he started to drive alone in his car from New York to California. In Nevada he met two nice-looking young men who asked him for a lift. They said they were on their way to college. Mac never refused a kindness to any human being. Of course he agreed. Near Lovelock they suggested that it would be well to fill his water bottle from a spring beside the road. While he sat in the car one of the men doped the canteen and offered him a drink. A short time later the drugged water made him so sleepy that he could not go on.

"I guess I've got a touch of the sun," said he. "I'll have to draw over to the side of the road and sleep awhile."

"We're sorry, but never mind. It is only three miles to Lovelock; we'll walk. Thanks a lot for the ride."

Mac got out and extended his hand to say good-by to one of the boys. The other stepped behind him—and that was the last he knew. Hours later he was found beside the road unconscious and was taken to a hospital. His money was gone and there was a great bruise on his head.

We have been able to get only fragments of what happened after that. For three days he remained in the hospital and left much against the doctor's wishes. Twice we heard of him from friends in Denver and San Francisco, each one saying that he complained of unendurable headaches. On September 3rd his body was found in his car in a lonely lane near Eureka, California, with a bullet from his own gun in the back of his head. He had been dead three days.

Murder or suicide? The coroner reported it to be the latter. I was in China at the time but I obtained a copy of the proceedings after considerable difficulty, and it certainly was a most casual inquiry. Just how a man can shoot himself in the back of the head and still have his revolver tightly clutched in his hand and all perfectly arranged, I fail to see.

Unless Mac was temporarily insane from the pain in his head I know that he never would have taken his own life. He was only thirty-seven years old and had everything to live for. Mac always laughed at life, taking the good with the bad, the thick with the thin; never complaining when Fate played him a scurvy trick. Kind-hearted, generous to a fault, loyal, affectionate, sympathetic, faithful to his friendships—that was J. McKenzie Young. His place in the Expedition never can be filled. When we return to Mongolia something vital which we loved will be gone from our life on the desert.

CHAPTER VIII

War in Peking

IN THE first half of this book I have discussed in general terms modern intensive exploration—the exploration of today and of tomorrow. The remaining chapters give a concrete example of such exploration as we conducted it in the Gobi Desert during 1928 and 1930. Our work was carried on with a background of war, banditry and political intrigue which made it exceedingly difficult. The details doubtless will not be the same in other countries, but during the present disturbed condition of world affairs, political barriers are almost certain to present themselves wherever the modern explorer wishes to work. The leader of a big expedition will usually find that overcoming diplomatic obstacles is infinitely more difficult and more nerve racking than conducting the field work. However, it is all a part of the job and cannot be escaped.

And now to take up the story of the Central Asiatic Expeditions where it was ended in "On the Trail of Ancient Man."[1]

When the Expedition returned to Peking from the Gobi Desert in the autumn of 1925, the scientific staff dispersed to various parts of the world. Doctors Granger and Nelson prepared for a winter along the Yangtze River; the former to continue his studies in the fossil pits near

[1] G. P. Putnam's Sons, 1926.

Wanhsien, Szechuan; the latter to work along the river banks examining the numerous caves which we hoped would give evidence of the occupation of primitive man. George Olsen fitted out a laboratory at headquarters in which to prepare the fossil collections, with a staff of native assistants. It was imperative that I should return to America to obtain additional financial support for the Expedition and to stimulate public interest by lectures and writing. Affairs in Peking were left in charge of J. McKenzie Young, assisted by Norman Lovell.

The 1925 Expedition, when we had a foreign staff of fourteen men in addition to twenty-six natives, Chinese and Mongol, had shown me that it was too large. Our mobility was sacrificed and it became unwieldy for the country in which we were operating. It required too much gasoline to move, and too much food to maintain the party. I had come to the conclusion that a foreign staff of ten was the maximum number that we could use effectively. I intended to organize the 1926 Expedition on that basis.

Just after I had sailed for America in October, 1925, the ship's radio picked up news of a civil war which had started near Shanghai. It caused me no worry because Shanghai is a long way from Peking and I did not believe that the war would spread northward. Even if it did, the northern Chinese never fought in the winter. I thought that it would last for a few weeks and then be settled as usual in the Chinese manner, without much bloodshed.

But it was just that year when things began to change in China and tradition and "good form" in the conduct of wars were completely smashed. The trouble did spread to the north like a flame and lasted all winter. Some of the bitterest fighting was in December and January near Tientsin. The second breach of "good form" was that they killed a good many people. Thousands of wounded poured into the city, and the countryside was strewn with dead. The railroad transport of all North China was paralyzed; for weeks no train ran between Peking and Tientsin, although according to the 1900 Protocol the foreign Powers have the right to maintain communications from the capital to the sea. Food was expensive and difficult to obtain. It was not a question of how soon a thing could be done, but whether it could be done at all. Therefore McKenzie Young had his hands more than full with preparations for the spring Expedition.

To buy tons of rice and flour, hundreds of pounds of sugar, coffee, beans and other food supplies in Peking was impossible. They had to be purchased in Tientsin but that port is eighty miles from the capital and no trains were running. Four thousand gallons of gasoline had been specially packed for us by the Standard Oil Company but, with the food, it was all in Tientsin.

Just about the time everything was ready, renewed hostilities began. Gradually the "People's Army" of the so-called "Christian General," Feng Yu-hsiang, were pushed back toward Peking. The only communication with Tientsin was by motors which were allowed to pass through the lines when there was no fighting actually on the road.

Since our caravan must start in March at the latest, Young decided to bring the gas and supplies up to Peking by motor. It was a laborious and very expensive method, but there was no other way. For three weeks Norman Lovell made a round trip every day. Leaving Tientsin early in the morning, he drove the eighty miles over what is only by courtesy called a motor road, and arrived in Peking at noon. While he was eating luncheon the car was unloaded and he was off again on the return trip to Tientsin before one o'clock.

During these days of transport he had many exciting experiences. The road swarmed with soldiers of one side or the other and a more annoying collection of uniformed brigands would be difficult to find anywhere. He carried a large American flag on the car but this did not prevent his being fired on time after time. There was a "no man's land" of ten or twelve miles between the opposing lines, and entering or leaving this area usually was attended with a good deal of danger. The advance and rear guards in the trenches not infrequently took a flying shot at the car just to see if they could hit it. Sometimes there was genuine suspicion that it might contain a machine gun party of the opposite side and the treatment Lovell received was distinctly unpleasant. Of course he would be stopped and questioned every few miles, but cigarettes and a plentiful supply of calling cards usually were sufficient to get him through.

Just before I reached China in March four new cars for the Expedition arrived from Detroit, for our fleet was to consist of eight Dodge Brothers cars. Lovell with three Chinese chauffeurs drove them to Peking. Unfortunately they happened to start on the day that the so-called "People's Army" were retreating along the motor road. Thousands of troops and lines of carts blocked the way. Our cars were heavily loaded with gasoline but the soldiers literally took possession of them. So many men piled upon each car that they could hardly pull even in low gear. The springs were absolutely flat and had they not been designed for especially rough work they certainly must have been ruined. The soldiers were very ugly and the fact that the cars bore American flags and that the road was a recognized avenue of traffic for autobiles made not the slightest difference. Lovell's feat in getting our gasoline and supplies to Peking is in itself sufficient testimony to his courage and tact.

It was the first week in March before our equipment was assembled at the Expedition headquarters in Peking and the caravan should have left by mid-February. When the first war rumors developed in the early winter Young had sent word to Merin, the Mongol caravan leader, to take our camels far out into the desert where they would be safe. Feng Yu-hsiang was confiscating every cart, mule, horse and camel within a hundred-mile radius of Kalgan, and the American flag which floated over Merin's tent would not have protected our camels from the soldiers of the "Christian General," or any others.

Meantime all the railroad cars on the Peking-Suiyuan Railroad had been seized by the military. It was virtually impossible to move a pound of freight to Kalgan even by paying the most exorbitant "squeeze" for the use of a freight car. Sometimes a merchant did manage to get one started from Peking, but it was usually side-tracked at Nankou, halfway to Kalgan, there to remain indefinitely.

Young had exhausted every means to get our supplies to Kalgan, without success, when I arrived in Tientsin on March 27, 1926, with J. B. Shackelford, the Expedition photographer. There had been sharp fighting not far from Peking and the authorities were loath to allow Young to drive to Tientsin to bring us up. His trip down was comparatively uneventful for those exciting days, although he was halted and questioned a dozen times by soldiers of both sides. When we returned two days later the road had been mined in fourteen places by Chang Tso-ling, as preparation for a counterattack by the "People's Army." There were thousands of soldiers but we had no trouble although we became a bit jumpy when driving over the mines which we hoped were buried deeply enough not to explode from the impact of our motor.

A week after our arrival in Peking I managed to get two freight cars for Kalgan through the influence of Mr. C. S. Liu, Director General of Railways. He sent word at four o'clock in the afternoon that they were available but said that we must load them at once or they would be confiscated by the soldiers. We worked a good part of the night taking over the gas and supplies. One was a covered steel car loopholed and bullet-marked from its use as a machine gun nest; the other was a flat car upon which we drove two motors and piled them about with cases. Young and Shackelford chaperoned the things to Kalgan without serious incident and returned to Peking. Word had been sent to Reverend Joel Eriksson, a Swedish missionary who acted as our agent in Mongolia, to bring in the camels, and as soon as they arrived I expected to go with Young to see them started.

In the meantime Peking had been having daily visitations from an aëroplane. Every morning promptly at ten o'clock the plane sailed out of the south, dropped a few bombs on the city and flew back again to Chang Tso-lin's lines. The roof of the Peking Hotel was the best place from which to see the show. Immediately "Bombing Breakfast" became the newest social diversion. A dozen guests would be invited to breakfast in the hotel at nine o'clock. At five minutes to ten they would adjourn to the roof, watch the plane do its stuff and then jump into motor cars to inspect the "scene of devastation." As they were small bombs filled with black powder the damage was slight. Sometimes a coolie or two would be a casualty. They never killed any soldiers.

However, they did have the American Legation staff pretty worried because it is situated very close to the Chen Men Railroad station. Several times the bombers tried to drop their "eggs" on the trains and missed by many hundreds of yards. Finally a message was sent to Chang Tso-lin to please cut it out, and he ordered his airmen to transfer their attentions to other parts of the city.

On April 12th I gave a bombing breakfast myself and was disgusted because the plane did not appear at the usual hour. It was the only day that it had missed and I felt like sending an official protest to Chang at having spoiled my party. At eleven o'clock I went with my Number One Boy, Lo, to the Hsichihmen railway station in the northwest city to make arrangements for sending the motors of our Expedition to Mongolia. Just as I drove into the broad plaza in front of the station a plane roared overhead. The pedestrians scattered, but a company of soldiers just marching out did not even glance up. Suddenly a bomb landed with a terrific crash thirty yards to the right of us on the other side of a high mud wall. I stepped on the gas, hoping to get into the heavily roofed station for protection, but a second bomb landed just in front of the car. Since we were both going in the same direction I decided to let the plane win the race.

Jumping out of the car I made for an armored train standing on the tracks. My Chinese Boy, Lo, flattened himself against a wall. With a dozen Chinese I crawled under the train and stretched out between the steel wheels parallel with the axle. Suddenly a bomb exploded with a deafening report not fifteen feet from my shelter. The iron fragments "pinged" against the car wheels like rain and I never knew how small I could make myself until that moment. A few seconds later two others crashed on the opposite side of the train. One iron slug came in at an angle and buried itself within two inches of my face. I dug it out and burned my fingers nicely as it was red-hot.

All was quiet for a few minutes and I crawled out, thinking that the raid was ended. But the plane had only circled and was directly above us, very low. It was a small four-seater carrying two men. One of them was chucking out the bombs by hand over the side. Before I could duck back under the train a bomb exploded a few feet away, then another on the other side of the train.

The place was a mess of flying gravel, chunks of iron and clouds of white smoke. In the midst of it a Chinese woman wandered out into the open space. I yelled at her to come under the train but she seemed dazed and wouldn't move. Just then the plane zoomed down again and dropped a bomb right in front of her. Her entire head suddenly disappeared as though it had been cut off by a knife.

My Chinese Boy at last decided to obey my calls to come to the train. He had just scrambled in beside me when another bomb killed the four coolies who had been standing with him against the wall. The airmen evidently were trying to get that particular train for they dropped fourteen bombs within a few yards of it. But we were pretty well protected, for even if he had landed a direct hit, the bombs were too small to have wrecked the steel cars.

We were having rather a lively time what with the noise and smoke and the groans of a dozen people who had been injured. Just beside a huge tank of the Standard Oil Company, which contained twenty-four thousand gallons of petroleum, was a primary school. A bomb went through the roof killing or injuring forty children. If the tank had been hit we would have had a terrible conflagration.

Finally the plane left and I went up to the station. We were no more than inside when the airmen returned and gave us another deluge. The vicinity of the railroad was such an unhealthy place that Lo and I made a dash for safety in the motor. It so happened that that was the only raid during which any considerable number of people were killed and that I was the only foreigner who ever was in real danger from the planes. The next day the foreign Legations sent emphatic protests to Chang Tso-lin and the raids ceased.

Not long after my bombing experience I got mixed up in the war again and pretty seriously too. This time it was with three members of the Central Asiatic Expeditions, Shackelford, Hill and Beckwith, and a captain of the U. S. Marine Corps.

The night before I was dining with the American Minister when firing began just outside the city and we all adjourned to the roof of the Peking Hotel to watch the show. Machine guns showed in a steady stream of light along the southern horizon, punctuated by the wide flashes of heavy guns. The American Military Attaché told us that Feng, who held Peking, had begun a new offensive and might even push Chang's army back to Tientsin. But the usual thing happened. One of Feng's generals was bought off by the opposing side and the advance became a retreat.

I had to get through the next day to Tientsin and thought that an American flag would give us immunity from fire. The gates of Peking were heavily guarded but the soldiers let us pass. Carts were already coming into the city loaded with grain, camp gear and soldiers. Cavalry streamed by and then thousands upon thousands of infantry. They were retiring in good order and seemed most cheerful. An officer told me that Chang Tso-lin's troops had taken Tungchow, fourteen miles from Peking and were looting the city but that there was no fighting.

We drove on slowly and eventually passed beyond the rear of the retreating army. For three or four miles the countryside was deserted, houses closed and all as quiet as the grave. We were five or six hundred yards from the ancient marble bridge at Tungchow when there was a sharp report and a pebble flew from beside the front wheel.

"That was a shot," said Shackelford.

"I don't think it was," I answered.

"Don't you fool yourself," said Hill. "It was a shot all right."

A second later the argument was settled. We had just rounded a curve and were in sight of the marble bridge. It was surrounded by a mass of soldiers with a machine gun in their midst. As our car came in sight the gun opened up. The bullets were kicking up the dust just in front of us but it was aimed too low. The soldiers could see the American flag plainly enough but that made not the slightest difference.

Fortunately at this particular spot the road was wide enough for the car to be turned and I backed around in record time.

"Every one down in the car," I yelled.

Shack, who was in the front seat with me, got his head down but only inverted himself, for he is rather fat, and his rear extremity was completely exposed. He was just like the proverbial ostrich. The soldiers had elevated the gun and the bullets were buzzing like a swarm of bees above our heads.

Forty yards down the road a sharp curve took us out of sight of the machine gun. The other men crouched in the bottom of the car. Since I was driving I could see all the fun. It was a pretty rough road but the speedometer showed fifty miles an hour as we went back. The ride became an exciting one. All the houses which had seemed so peaceful actually were occupied by the advance guards of Chang Tso-lin's soldiers. They had let us pass because of the American flag, but when they heard the firing in our rear and saw us returning at such a mad speed, they evidently thought that we were anybody's game.

For three miles we ran the gauntlet of firing from both sides of the road. I would see a soldier standing with his rifle at the ready waiting until we came opposite. Then "bang" he would let us have it. Sometimes they fired in squads; sometimes singly. Most of them merely pointed their rifles at the car and the bullets struck just behind us. Two or three zipped in between my head and the windshield.

There was a small temple just ahead and I saw a villainous looking fellow standing at the ready waiting for us. Something seemed to tell me that he was dangerous. He brought his sights into line and the muzzle on me.

"This one," I thought, "will probably get me."

My tummy took a nose dive. I ducked my head just as he fired and his bullet went through the brim of my hat. That was the closest call any of us had. I was afraid that one of the tires would be hit. At that speed we'd have turned over as sure as fate. But there was nothing to be done about it. I really had the best job because the others couldn't see what was going on, and driving kept me busy.

Soon we approached three of Feng's soldiers—the last men of his rear guard.

"Perhaps it'll be safer for us if we stop and pick these men up," I suggested.

My companions agreed, and when we drew alongside

the soldiers we gave them a lift, letting them stand on the running board. But as we approached Feng's main body, our men became panicky, probably because they knew that a summary court-martial awaited them if their officers discovered that they had deserted their rear guard position. With our car going at twenty-mile speed, and without warning, one of them stepped off backwards.

He fell to the road and a wheel ran over his hand. I stopped the car. The coarse gravel had acted like a grindstone, shredding his hand horribly. I put on a hasty tourniquet and offered to take him in to first aid, but he waved me away frantically. No officer would catch him!

We went on, and soon had to slow up because of the straggling army of men. Soldiers began to jump on the car. In spite of my protests, more and more climbed on until we actually carried eighteen clinging men. Both running boards were jammed solid, others hung on the rear, two sat astraddle the hood, and their rifles were piled on the top. I couldn't see to drive. The car would barely crawl in low. Vainly I complained. These men spoke the Shantung dialect, which is difficult to understand. Then came the accident that set off the fireworks. One of the men precariously perched on the front of the hood fell off. A wheel ran over his leg and the heavy load plus the gravel mangled it badly.

The Chinese have a great tendency to talk themselves into a rage. They yanked all of us out of the car, crowded close, and shouted and gesticulated themselves into hysterical anger. Finally, with their tempers whipped to a white heat, they lined us up against the car and cocked their rifles. Things looked pretty bad.

Just then an officer appeared. Fortunately, he could speak Mandarin Chinese and I explained what had happened.

He said, "I am a staff officer. I can't control these men. You must get off the road at once or you'll be killed. Drive down the bank there into the fields. I'll stay here until you are out of rifle shot."

It was a difficult job to navigate over the plowed ground, but somehow we got to the gate of Peking and into the city. The experience affected each of us differently. I had been so busy driving that there was no time to be scared; or at least not to give up to the feeling. I had got the other fellows into the jam and had to get them out. It was not until we were back in Peking that the reaction caught me. We had driven to the Club for all of us badly felt the need of a strong drink. My hands were shaking so that I actually couldn't hold a glass. I was trembling all over and felt awfully weak and sick.

One of the other men who lived with me had been perfectly cool throughout the entire performance and afterward. At two o'clock the next morning he came to my room in violent hysterics. The nerves which he had kept under rigid control had suddenly snapped and for two hours I had a beautiful time getting him back to normal.

CHAPTER IX

Attempts to Reach Mongolia

THE day after our experience in trying to reach Tientsin, I decided to join Mac Young, who had gone to Kalgan. At eight-thirty in the morning when we drove to the Hsichihmen we found the gate closed and sandbagged. The soldiers said that no one could go out or enter the city. We made the rounds of the other gates and found them all heavily fortified. The streets were deserted, the shops closed and a strange air of preparation for a great calamity pervaded the city. If the Fengtien troops forced the gates, Peking would certainly be looted. Every Chinese of political importance who could find a lodging had fled to the Legation Quarter. Most of these same men had been loudly demanding the abolition of Extra-territoriality and the dismissal of the Legation Guards but a few days earlier; yet at the first hint of danger they dashed to the foreign Concessions as the only place of safety! This is what always happens.

A good many foreigners who had cottages at the race course, seven miles west of the city, were caught outside the walls and had to remain there for several days in the midst of roving bands of soldiery. The first Secretary and the Counselor of the British Legation tried in vain to get the guards to open the gates. At last they found a part of the wall where loose bricks had been removed by smugglers. A number of Chinese were on top and when the foreigners produced a silver dollar a rope magically made its appearance. The diplomats were hauled unceremoniously up the wall losing a certain amount of "face" but at least getting to their posts in the Legation.

Notices were sent to all foreign residents in Peking by their respective Legations instructing them where to assemble in the event of extreme danger; green lights and guns were the designated signals. From the concentration points the foreigners would be escorted to the fortified Quarter by armed guards.

We had a carefully thought-out plan for the protection of the Expedition headquarters and the great quantity of valuable equipment in the compound. With machine guns posted on the roofs we would be able to present a pretty strong defense against any looters or even well-armed soldiery. All the men offered to remain with me and protect the house if necessary.

The "City Fathers" did some sterling work in persuading the Fengtien generals not to let their troops come into the city, urging that foreign complications would arise with the inevitable looting. The soldiers encircled the walls but with the gates shut and sandbagged not a man was allowed inside.

In the meantime the social life of Peking proceeded much as usual. The only fish we could get came from the lake in the Forbidden City, fruit was non-existent, and there was a shortage of fresh meat. However, nothing is allowed to interfere with dinners and dances, polo and tennis.

The third evening of our siege I was amazed at the

arrival of Mr. and Mrs. W. Douglas Burden of New York. They had reached Tientsin a few days earlier and I telegraphed them to wait there until it was possible to bring them up. They had discovered that two men in a motor car were attempting to get through to Peking and took the chance of coming with them. Fortunately for them they picked up on the road the Fengtien general in command of the air forces whose car had broken down. He was one of the few men for whom the city gates could be opened.

I was at the Chi Hua Men talking to the soldiers when they asked me to please draw my car to one side as a very important general was arriving. The sandbags were being removed from one side of the gate and when it swung open just enough to admit a car what was my amazement to see my friends come in!

They had had an adventurous trip up from Tientsin. The human heads hanging from posts along the road and the evil-looking soldiery impressed them with the fact that their own lives were by no means safe. Theirs was the first car to come over the road and Peking was keen to learn what was happening outside the city.

After a few days we discovered that Feng Yu-hsiang's army had retired up the railroad and entrenched themselves at Nankou, the pass through the Great Wall on the way to Kalgan. For many hundreds of years this pass acted as a strategic point protecting Peking from invasions of the Mongols and Tartars from the north. Now it acted in the opposite way by preventing the Fengtien army from following Feng's so-called "Christian soldiers" to their headquarters at Kalgan.

I had had no news from Young since he reached Kalgan on the last train that ran from Peking. Telegrams could not be sent and although wireless messages were accepted by the Chinese station in the Temple of Heaven they never were delivered. I tried every possible means of communication without success. Finally the American Consul in Kalgan got a radio through to the Consul General in Tientsin, who forwarded it to the Legation.

It said that our caravan had finally started on April 26th, after having been commandeered three times by the soldiers in defiance of the permit given by the General in command at Kalgan. He had lost so much "face" over the matter that he furnished a military guard to see them beyond the limit of soldier activities. Young stated that he had tried to get down to Peking but had been turned back by the troops.

The days dragged on interminably for all of us. All the staff were assembled with the exception of Young, who was in Kalgan. The cars were loaded in the courtyard of the headquarters, and the Expedition ready to leave at an hour's notice, but we were effectually blocked. Although the Fengtien troops were not pushing their advance against Nankou, heavy artillery fire could be heard every night in Peking.

I tried to obtain permission to go by way of Shansi and reach the Mongolian plateau west of Kalgan. Just when it seemed that it might be arranged, Feng's army advanced along the proposed route. Therefore our plans came to an abrupt halt. There was a lull in the fighting at the Nankou pass, and I entered into negotiations for our passage through the lines. But the Generals said that as there was a good deal of guerilla warfare going on we would certainly be killed if an attempt were made.

About two weeks later I was amazed to have a telephone message from Young. He had just arrived in Peking from Kalgan after a strenuous and adventurous trip. At the hotel I found him unshaven, gaunt and hollow-eyed. He had come by way of Shansi and had traveled about six hundred miles in order to get to Peking. By the direct Nankou route Peking is only one hundred and twenty-four miles from Kalgan.

Young had walked a good part of the way and the only food he could obtain along the road was a few boiled and salted eggs. We got him into bed and after a good many hours of sleep and some decent food he was fit and ready to do it over again if necessary. He reported that food in Kalgan was very low. For weeks the foreign residents had had no coffee, tea, sugar, milk or butter, and but very little flour. Occasionally they could get fresh mutton. Cigarettes were entirely gone. Fortunately the British American Tobacco Company had a large garden in their compound and by forcing the vegetables under glass the foreigners would be able to carry on for some time longer. Nevertheless the situation was serious, for typhus had broken out and the only doctor in town had died of the disease.

Although Young had been able to get through both lines in Shansi it was his opinion that it would be suicide to attempt it with six motors and a large party. We were certain, he said, to be annihilated by snipers even if we had permits and carried American flags. It wasn't good enough.

Only one other possible route to the plateau remained: that was by way of Jehol, the old summer residence of the Manchu emperors, then over to Dolon Nor in Inner Mongolia, and down to Kalgan. Jehol is one hundred and forty miles from Peking but the road is unspeakably bad. Lovell and I went in one of the Expedition cars and I enjoyed the trip for it took us through some of the most beautiful scenery in North China. A few miles beyond the An-Tung Men, the north gate of Peking, we saw several thousand Fengtien cavalry coming across country at a sharp trot. They gave the appearance of retreating soldiers, but we were not molested by them.

Jehol itself is a beautiful spot and I was looking forward to enjoying the old palace and the tombs. But we found that the place was literally swarming with soldiers who were so obnoxious that we remained there only a few hours. I had a letter of introduction from the American Minister to the Military Governor and, after a considerable wait, was given an interview. He was a former bandit but received me courteously in a small room at the extreme right of the old palace. The walls above the *kang*, or bed platform, on which he sat, were hung about with automatic pistols. There were ten of them and one was always within his reach.

He would not even consider the suggestion of letting us through to Dolon Nor. He said that, were he to do so, he would be signing our death warrant, for the region was so infested with brigands and deserting soldiers that it would require a very strong armed guard even to take us through his own lines. Doubtless he was right, for in his outer office I had talked with two of his tax collectors who had been fired on by brigands while on their way to a station a few miles from Jehol. In the city we visited a British missionary and his wife, the only foreign residents. They said that for months they had been in a state of continual unrest due to the soldiers who swarmed through the place. Both of them seemed to be pretty well shaken nervously, and I can well imagine what they had been through.

When Lovell and I returned to Peking we had an unpleasant experience just outside the north gate. Although but few soldiers had been there when we went out two days earlier, the right wing of the Fengtian troops had retreated from the Nankou pass and were entrenching themselves across the road.

In spite of our American flag we were stopped by soldiers who treated us as though we were bandits. Our passports were quite in order but unfortunately for us the sergeant in charge could not read Chinese. For an hour this wretched guard kept us covered with loaded and cocked rifles while they searched the car and ourselves. In spite of my protests, the soldiers would neither take us to an officer who was able to read his own language, nor allow us to go. At the slightest motion on our part the boy soldiers (they were about sixteen years old) would throw their rifles to their shoulders. I indicated to our captor that the rifles might easily go off accidentally in which case we would be killed. His only reply was "*Mayo fadzu*" (It can't be helped).

At last an officer who had enough education to read Chinese appeared. As soon as he had examined our passports we were allowed to proceed. Had we been killed even accidentally, they would have reported that we had attacked them and they had been forced to shoot in self-defense. The soldiers showed quite plainly that they had spent a most pleasant hour in annoying two foreigners who, in front of cocked rifles, were helpless.

Two other attempts to reach the plateau failed. There was always the hope that one of those sudden changes so frequent in Chinese politics would open the way for us, and it was not until June that I had to finally admit that the 1926 Expedition must be abandoned.

Our caravan had gone as far as Shara Murun, two hundred fifty miles out on the plateau, and there had awaited word from us. Hearing nothing they returned to Hallong Oso, one hundred miles north of Kalgan. One of the Mongols, Tserin, came in to learn what had become of us. He arrived just at the time when food in Kalgan was almost exhausted and the eighteen foreigners were facing real hardship. The American Consul managed to get a radio message to the Legation asking if they might take over our food. I immediately agreed and the supplies were brought to the city. The besieged foreigners sent us an enthusiastic note of thanks.

It was not until mid-August that Feng Yu-hsiang's troops abandoned the defense of Nankou and retreated into Shansi. The "Christian General" fled to Russia via Urga where he at once renounced all his Christian doctrines. Chang Tso-lin set himself up in Peking as Dictator of North China.

I had long wished to make a reconnaissance of Yunnan Province from the standpoint of archæology and palæontology. As conditions appeared to be normally quiet there, Nelson and Granger prepared to spend the winter in that beautiful province. From the experience I had gained on my 1916-17 Expedition, I was able to map out for them a tentative program and they left in August for Yunnanfu.

Their winter in Szechuan had been interesting and valuable. Granger obtained a splendid collection of Pliocene mammals from the fossil pits at Wanhsien, which supplemented his former work. Nelson was disappointed in finding no traces of Palæolithic man in the caves along the river. Most of them had rock bottoms. He came to the conclusion that the river had not been used as a highway of travel until man had advanced far enough to learn the use of boats. He did, however, obtain a considerable collection of Neolithic implements and of those representing the interesting pre-Chinese culture first discovered by Dr. J. G. Andersson.

As there was little that I could do in Peking during the winter, I sailed from Shanghai on September first for America. Our fruitless summer had cost considerably more than would have been expended during a season's field work, and more money was urgently needed. I had also accepted an invitation to present the results of our explorations before the Royal Geographic Society in London at the Asia Lecture on November 10th. My winter was a busy one and added some fifty thousand dollars to the treasury of the Expedition.

CHAPTER X

Politics and People

DURING the winter of 1926-27 important political events took place in China. The British Concession in Hankow was forcibly taken by Chinese soldiers. To the stupefaction of all foreigners in China, the British Government not only made no attempt to reclaim it, but eventually officially returned it to China.

Chiang Kai-shih, in command of the southern army, was energetically making plans to attack the north with the object of bringing the entire country under the control of his party. He was ably assisted by the Russian, Borodin, and his army was directed by Russian generals. It had remarkable success and eventually moved on Shanghai with the openly avowed intention of taking that city, ousting the foreigners and claiming the port for the Chinese. Flushed by their entirely unexpected success in the Hankow affair, the Chinese had decided that the Foreign Powers would not protect their interests and that if they used force they could drive all foreigners out of the country. They demanded the return of the Tientsin concessions and of the Legation Quarter in Peking. Propagandists, directed by Borodin and his Bolshevik organization, were active in all parts of China and particularly in the northern armies. The "battles" which Chiang Kai-shih was heralded all over the world to have won against the northerners were farcical. Propaganda had so thoroughly done its work that usually the enemy retreated upon the appearance of the southerners, or else deserted to their side.

But the Foreign Powers had been driven too far by the advance upon Shanghai, and war vessels and troops began to arrive from all quarters of the globe. The British sent battleships and a large force which were first upon the scene. It is generally admitted that their prompt action saved a most horrible wholesale massacre of foreigners.

An indication of what would have happened all over China was given at Nanking. A southern army marched into the city, killed several foreigners, attacked the Consulates of two or three foreign powers, and began a systematic looting of every foreign house. Eventually the foreigners realized that they would be murdered to the last child and gathered on a hill belonging to the Standard Oil Company, where they were besieged by the murderous soldiers. A gallant American marine ascended to the roof of the building and while bullets were spattering all about him signaled to the warships in the river. The officer in command of the American destroyer said, "I'll either get a medal or a court-martial for this, but here she goes." His ship laid down a box barrage about the hill. As the first high explosive shell dropped near the house the Chinese soldiers ran pell-mell in every direction. The foreigners were rescued by a landing party when the place had been cleared of the Chinese murderers.

Meanwhile thousands of foreign troops had gathered

in Shanghai. Barbed wire entanglements were erected about the foreign concessions and the city put under martial law. A curfew kept all residents indoors after ten o'clock in the evening. One or two clashes took place with considerable loss to the Chinese, but no determined attack was launched against the concessions. The northern forces, completely disrupted by Bolshevik propaganda as they were, retreated, and the southerners occupied the native city of Shanghai.

All foreign legations had ordered their nationals from the interior of China for it was quite evident that the Chinese intended to repeat the Boxer attempt of 1900 and kill or drive out every foreigner. Reports were continually arriving of murders and outrages committed upon foreigners in various parts of the country.

Such in brief was the situation when I reached Peking in early April, 1927. I came via Korea by train. Every ship out of China and every train was packed with departing residents, but coming to Peking I was the only foreigner on the entire train.

At Tangku, not far from Tientsin, what was my surprise to see Granger and Nelson on the platform. They were just returning from Yunnan and we all came on to Peking together. Granger reported that their trip had been somewhat disappointing. They had at first gone south and then east of Yunnanfu because the western part of the province was so infested by brigands that the authorities would not allow them to enter it. They had discovered only one important fossil deposit; that was of Pleistocene age, but bandits drove them out after they had spent only three days there. Granger believed it to have most important possibilities and hoped to return for a further exploration. Nelson had fared little better. He had discovered no traces of Palæolithic man but did find further evidences of the Yangtze River Neolithic and pre-Chinese culture. I was sorry that they had not been able to explore the country to the west, for I believe that it will prove to be well worth investigating.

Upon reaching Peking we found that the foreign residents were thoroughly frightened. It was the first time that I had seen anything like a panic. Even the year before when the gates of Peking were closed and sandbagged, and Chang Tso-lin's wild Manchu hordes were looting and burning the countryside, few foreigners in the capital were even nervous. But the Hankow and Nanking outrages had inflamed the anti-foreign feeling which exists in the hearts of most Chinese. The poorly veiled hostility to foreigners by all classes made us realize that wholesale slaughter had only been averted by the arrival of foreign troops in China. The southerners were pushing slowly northward and the Legations were already advising their nationals to leave Peking. Efforts were made to get as many of the women and children as possible to go to Dairen, Japan, or Manila.

Suddenly the situation was completely changed by a dramatic raid by Chang Tso-lin's soldiers on the premises of the Russian Dal Bank next to the Soviet Embassy. As Chinese soldiers are not allowed to enter the Legation Quarter, the raid was made with the permission of the Diplomatic Body. Although they had agreed only to allow the soldiers to search the Dal Bank, they went much further and ransacked the office of the Military Attaché.

The diplomats protested, of course, but only formally. The raid took place at eleven o'clock in the morning. I happened to be at the National City Bank on the opposite side of the street and witnessed the entire proceeding. It was most dramatic and totallv unexpected by the Russians.

Until six o'clock that evening the searching proceeded. Many Chinese and Russian propagandists were unceremoniously hauled out from their hiding places and hurried off to jail. One Bolshevik attempted to burn various important documents, but the fire was extinguished before many had been consumed.

Even in his wildest dreams Chang Tso-lin could hardly have believed that his raid would produce such important results. Evidently the Bolsheviks had completely depended upon the diplomatic immunity of the Embassy and had used it as a central clearing-house from which operations were conducted in various parts of the world. Most incriminating documents were found. Few people realize that raids which subsequently took place in London, Paris and the Argentine were made upon information obtained at the Soviet Embassy in Peking!

Chang Tso-lin then set to work systematically to rid North China of the propagandists. Those Chinese who were caught in the Embassy raid were slowly strangled. Search parties were busy day and night rounding up the others whose identity had been disclosed by the captured papers. Hardly a day passed that one or more persons were not executed at the public ground opposite the Temple of Heaven. The place has the appropriate name of "The Heaven's Bridge." The propagandists fled from

North China like rats deserting a sinking ship. In a very few days Peking and its environs had lost its apprehension and settled down into quite a normal existence.

Heavy fighting was going on but it was still some distance from Peking and the raid had so disrupted the propaganda system of the southerners that Chang Tso-lin made some progress in driving them back. Finally, however, they recovered from the blow to their plans and again began to advance slowly northward.

The prospect for continuing our exploration in Mongolia could not have been blacker. Even had I been able to get the Expedition away from Peking, the American Minister would have prohibited us from leaving. Hardly a single foreigner was left anywhere away from the seaports of China. Still, it was most disheartening to face another season of inactivity. Granger, Olsen and Nelson sailed for America. Only McKenzie Young remained with me at the headquarters. We proceeded to liquidate certain effects of the Expedition, put others in a place of comparative safety, and reduce current expenses to the minimum. I decided to stay in Peking during the winter, hoping that one of those sudden changes that so frequently occur in Chinese politics would give some encouragement for an expedition in 1928.

Spending a winter in Peking can hardly be called one of the hardships of an explorer's life. It is the most interesting and the most delightful city in the world and I say that advisedly. I lived in a beautiful old Manchu palace; had a staff of eighteen efficient servants; a stable of polo ponies and hunters; and a host of friends. Peking is the one place left in the world where one can live an Arabian

Nights' existence. One rubs the lamp and things happen. Don't inquire *how* they happen; just rub the lamp!

Peking with its history, its temples and palaces, is fascinating as a city; the Chinese themselves are interesting and the foreign community is always amusing. It is a cosmopolitan group of the most varied sort. It would require a real effort to give a dinner of only one nationality. Usually as I looked over my own table I would see five or six, and once out of fourteen people there were nine nationalities. It just happened that way. They were the people I liked—the outdoor set—those that rode, raced, hunted, played polo, and danced. The bridge and mahjongg crowd I saw less often for our interests were not the same.

Almost anything can happen in Peking. People do the most extraordinary things—things they never would do any other place on earth or even want to do. I suppose part of it is due to the dry exhilarating air which acts like a perpetual cocktail; part to the close association of races with radically different social and moral standards; and part to the subtle influence of the East. Kipling expressed it when he wrote:

Ship me somewhere east of Suez,
Where the best is like the worst;
Where there ain't no ten commandments,
And a man can raise a thirst.

The foreign social life centers about the legation, of course, and it is always interesting. The Diplomatic Body are the only ones who really take themselves seriously—

the rest of Peking laughs at everybody and everything. Next to the "D. B." our most fruitful source of amusement was likely to be our Congressional junketers, many of whom came to the East only for the ride. They were a joke to us who were not officials, but to the Legation staff it was tragic.

A certain United States Senator and his wife, who for obvious reasons shall be nameless and who with his wife was a laughingstock of his own colleagues, gave us the greatest kick of the year. Unfortunately for the poor Counselor of Legation, the Minister was absent and the former was Chargé d'Affaires. The Counselor was the proud owner of a Lincoln motor car. In it he met Senator and Mrs. Senator. On the way to the Legation the Senator remarked,

"I think it is an unjustifiable extravagance for the Government to furnish Lincoln cars for its representatives. A Ford is good enough for me. I'm going to do something about it."

"But this happens to be my own personal car," said the Counselor.

"Well, anyway it uses more gasoline than a Ford and I suppose the U. S. Government has to pay for that. Moreover, we are a democratic nation and cars like this create a bad impression."

So it went for three days—just one damn thing after another. The Counselor was nearly mad but he stuck it out and on the last night of the Senator's stay gave a beautiful dinner for them in his own home. A visiting Ambassador to Washington and the British Minister to China were among the guests. I happened to be there too.

According to diplomatic precedent, the Ambassador was seated on the right of the hostess, then came a lady and next to her the Senator. On Mrs. Counselor's left was the British Minister; then Mrs. Senator and then I.

The Senator refused to be seated because he was not placed at the hostess' right hand. There was an awkward pause while it was explained to him that an Ambassador and a Minister of Foreign powers outranked a Senator while dining in a United States Legation.

With extremely bad grace he finally sat down. Before the fish was served, my turn came. Mrs. Senator examined my place card.

"Hum, Doctor Andrews. Oh, I'm so glad to be seated next to a doctor. Do you know, almost as soon as we reached Japan I got dysentery. I've had the most awful time. I've just"—then followed such intimate details of her physical condition that I had to do something about it or have my dinner ruined.

"But I'm not that kind of a doctor, Mrs. Senator. It really doesn't do any good to tell me about it. I'd advise you to consult a physician."

"Not a doctor? Not a doctor?" said she indignantly. "Your card says 'Dr. Andrews.' What are you?"

"I'm a Doctor of Science," I replied meekly.

"Doctor of Science! I never heard of it. It is perfectly ridiculous to call yourself a doctor if you are *not* a doctor."

Then, thank God, she turned her back to me and gave the British Minister her attention for the rest of the evening.

After the women had gone into the drawing-room and

coffee was being served, the Senator remarked to our host,

"Pretty nice dinner. Good wine—I suppose the United States Government is paying for this?"

The Counselor had arrived at such a state that nothing mattered any more.

"No," said he politely. "I pay for it. But if you'd like to know what it cost I'll find out."

He called his Number One Boy.

"How much this dinner cost, Li, each person?"

"Each piecie, about two dollars, Master. This our number one dinner. Champagne cost eight dollars bottle. We drink four bottle. Fourteen people twenty-eight dollar. Wine, four bottle, thirty-two dollar. Altogether cost sixty dollar. Not too much."

All this in front of the Senator.

"The dinner, Senator, cost sixty dollars," said our host.

"Yes, but that is silver. Exchange about two for one. Thirty dollars. Not bad, not bad. Couldn't do it for that in Washington. But I'm glad the taxpayers don't foot the bill. Have to watch these things, you know."

For the honor of the United States, however, I must admit that this was a particularly bad specimen to turn loose on an unsuspecting world. We had some others that same summer who were really good eggs.

You just can't live happily in China unless you have a sense of humor. Otherwise you would be in a rage half the time. The Chinese don't think as we do in the first place. Their whole method of reasoning is so utterly different from ours that you have to accept the fact and try to adjust your mental perspective to theirs. After awhile you may achieve partial success but mostly you have to depend upon your sense of humor to pull you through without a nervous breakdown. Their logic is so different from ours.

Far be it from me to say which kind is better, but certainly theirs is more convenient. Witness an example at the former International Race Club in Peking. A mile race with only Chinese jockeys was the feature of the day. The judges were all Chinese. The favorite pony ridden by a popular jockey carried most of the money. But he was a nervous little beast and at the post broke away and ran the entire course. The starter held the field for he didn't dare let the others go without the favorite. Finally they all got off together with the favorite in the lead. But he couldn't quite stick the second mile and dropped back to finish a good second, only a length behind the leader. But did that bother the judges? Not at all! They declared the favorite the winner, much to the satisfaction of the crowd.

They argued, "Already that pony has run two miles and the others only one. He finished second. It is quite obvious that he is the best pony and would have come in first if he hadn't bolted."

Of course they might have been slightly influenced by the fact that one of the judges owned the favorite!

Another example of Chinese reasoning: I was driving back from the Western Hills one afternoon when my car ran over a duck crossing the road in front of a farmhouse. I stopped. The peasant's wife was wailing over the corpse which was obviously far beyond repair.

"Lao yeh, Lao yeh" (Honorable Master), she moaned, "You have killed my duck, my beautiful duck, the finest one of all my ducks."

"I'm sorry, but I'll pay, of course. Then you can buy another finer duck."

We bargained, and it was eventually agreed that sixty-five cents was a fair and proper price for careless ducks which got squashed under motor cars. The woman started back into the house with her money. Suddenly she whirled about.

"Oh, I have forgotten the egg. I must have ten coppers for the egg."

"What egg?" said I.

"The egg that's in the duck. Look! You'll see there is one there. She was just going across the road to lay an egg when you killed her."

I investigated and sure enough there was a broken egg in the *corpus delicti.*

"Do you want me to pay for all the eggs your duck would have laid this summer?" I asked.

"No, of course not. Those eggs don't exist. But this one does. There it is. You can see it. I could have sold it for ten coppers."

The reasoning was indisputable. She got the money.

I have discovered, too, that you can have a lot of fun at housekeeping in China if you don't let it get on your nerves. "Squeeze" drives foreigners mad at first until they come to realize that it is a custom of the country and that all you can do is to keep it within reasonable limits. Housekeeping becomes a game and you've got to play it with a sense of humor, realizing before you start that you can't win. Your servants try all sorts of things and usually they get away with them. They think so much further than you do. They plan a stunt to make a little extra money out of the master, and since he is completely unsuspecting, the thing is done before he knows what is going on.

My Number One Boy gave me a small bowl of goldfish at Christmas. I was pleased because they were very nice goldfish with long flowing tails. The Boy called my attention to all their good points and I discovered that he knew a lot about goldfish.

After a week the Boy suggested that he buy a larger bowl for me with some especially decorative goldfish—it would cost about fifteen dollars. The bowl did look very nice. But one by one the fish died, as they have a habit of doing, and very unobtrusively were replaced with others. My monthly bills began to show a regular charge for goldfish.

By that time, under the expert guidance of the Boy, I had become very much interested in goldfish. I began to talk about them at the Club and pose as rather an authority. It had come early spring when the Boy suggested that I go in for goldfish in a big way. An enormous earthenware bowl was to be purchased and stocked with the very finest goldfish to decorate the central court. It was done, the bill amounting to a hundred and fifty dollars.

A few days later I noticed a new coolie about the house. I said to him, "What do you do here?"

"I belong goldfish coolie," he answered.

His job, I discovered, was to clean the fish jar once a

week and to scatter the prepared food on the water. For these arduous labors I was paying him eight dollars a month (Mex.).

My fish account now began to reach alarming proportions. In spite of the efforts of the goldfish coolie, no sooner did I get an especially beautiful specimen than it died. I suddenly lost interest in goldfish and told the Boy that I was through. He was to sell the fish and the jar and bring me the money. This was forthwith done and I was presented with twenty dollars as proceeds of the sale. My Boy explained that just then the goldfish market was very bad.

Having rid myself of the fish I forgot about them. Nearly a month later I happened to go into the kitchen compound, and there under an apple tree was my jar filled with the most beautiful goldfish. I called the Boy.

"What the devil is this?" said I.

With a bland smile he answered, "I liked the goldfish so much that I bought them from Master."

I smelled a mouse then and did a little investigating. My discovery was as follows: my Boy was a breeder of goldfish—one of the best in the city. 'Way back at Christmas time he had conceived the idea of getting me interested in goldfish. Having done this, all the fish which he acquired for me were purchased from himself at an exorbitant price. The goldfish coolie was a man who owed him money. He gave him the job and took his salary of eight dollars a month as part liquidation of his debt. He had purchased back from me my jar and all the fish at about a seventh of what I had paid for them. Out of the deal he had made several hundred dollars.

This was all a perfectly legitimate operation according to the Chinese code. I had had my fun out of the goldfish and just because I had been done in the eye as to price was my own fault. The Boy had not the slightest hesitation in admitting the whole plan. The only thing I could do was to laugh about it.

Many people believe that foreigners who live in China eat only Chinese food. Nothing is further from the truth. I think Chinese are the best cooks in the world. Mine had worked for nineteen years in foreign legations before he came to me. He could cook any kind of food of any country and prepare it most deliciously. Every morning he would present the day's menu written in perfect French. Some days I would have Russian food—others Italian or perhaps Swedish. But always Chinese food at least once a week for I like it enormously.

But not the so-called "delicacies." I'm all against those. No chicken windpipes for me. Neither do I like fishes' stomachs nor ducks' feet. From camels' humps, ducks' tongues and bears' paws I remain aloof. Sea slugs are worst of all. I'm not just trying to be funny. These are really honest-to-goodness dishes which are likely to be presented to you at any Chinese banquet in Peking if the host is rich enough and is giving a really swagger dinner.

The Chinese dearly love strange and unusual dishes. Not as food really, but as an excuse to demonstrate the ability of the cook in concocting a sauce which would make a piece of rubber boot delicious. A chicken's windpipe tastes just as you would expect a chicken's windpipe would taste—like nothing at all. But, if you want to, you can suck on the windpipe to get the sauce. Shark's fin soup isn't bad—in fact, I rather enjoy it. Likewise bird's nest soup. The latter isn't made of sticks and straw. It comes from the beak of a little swallow. Two glands on either side of the throat secrete a gelatinous substance like glue. When the bird is impelled to build a home, it sticks a bit of the glue to a rock, draws it out and fastens it a few inches away. Back and forth it strings the glue which hardens and forms a tiny half basket.

The hundred-year-old eggs are seldom a hundred years old. The process of preparation is hastened by laying down the eggs in lime and leaving them for a year or two. The yolk turns black, and the white, light yellow. Dipped in soya bean sauce the egg isn't bad.

But what I've been telling you is only of the strange dishes with which a Chinese gentleman amuses his guests and "gives face" to his cook. The meat balls, pork and cabbage, eggs, chickens, ducks, fish and a hundred other dishes are so delicious that it makes me hungry even to write about them. In the north the Peking duck is important most of all. He really ought to be canonized. Great white fellows, with thick fat, and tender as a squab. They have restaurants in Peking where only ducks are served in a dozen different ways.

An amazing thing about a Chinese cook is the way he can serve a delicious dinner with the most primitive cooking arrangements. Many foreigners rent temples near Peking as country places. Mine was called the "Temple of the High Spirited Insects." There were only a couple of tiny charcoal stoves in the kitchen which a foreign cook wouldn't consider as stoves at all. Yet my cook could produce an eight-course dinner which would make your mouth water, all prepared in the temple.

It was at this same temple that I had a most uncomfortable experience in a Chinese grave. My shoulder had been broken in a steeplechase and my whole right side was done up in a plaster cast. Lunching at the Temple of the High Spirited Insects on a cold January day, I walked by myself across the fields late in the afternoon. A tiny path led toward a village a quarter of a mile away. Near it was a newly dug grave, deep, but narrow. Idly looking in, my foot slipped and down I went into the pit. I landed on my back, fitting the grave as though it had been measured for me.

I lay there looking up at the patch of sky already golden with sunset glow, laughing at myself. Then I tried to get out. Not a chance. I was wedged in as tightly as a sardine. My whole right side was powerless because of the plaster cast and I couldn't move an inch. The temperature was ten below zero. Already I was numb from cold and began to realize that it was far from being funny. If some one didn't get me out I'd freeze to death long before morning. It was already dusk and the Chinese don't move about much in the winter after dark. I did get a little amusement, however, out of thinking what a sensation I'd cause when the coolies came to finish the grave and found me in it. An hour gone and I had just reached the drowsy stage which precedes death by freezing. Then I heard voices—somebody walking along the path. I shouted, but it was a pretty feeble shout. The Chinese apparently didn't hear it, for they went right on chattering. Terrified that they would pass by, I let out a yell that stopped them in their tracks.

"I'm a foreigner; I've fallen in the grave; I'll give you money to get me out," I bawled in Chinese. "Money to get me out. I'm a foreigner."

Then startled grunts and a pattering of feet as they ran away. But I knew that they would return. Curiosity would bring them back or, if they were too frightened, they would go to the village and tell the story. Pretty soon voices sounded and I saw two faces peering down at me over the edge of the grave.

"I'm a foreigner; I've fallen in; I'll give you money to get me out," I called.

They were two old men, hanging on to each other, absolutely terrified. But the words "foreigner" and "money" had penetrated their fright as I hoped they would.

As soon as they discovered that I wasn't a devil or a spirit, they were all friendliness. I told them to get ropes and to hurry. Back to the village they trotted. In fifteen minutes a dozen men, women and children were at the grave-side with rope, lanterns and a rickshaw.

I was hoisted out eventually, but too cold to walk. The rickshaw carried me to the nearest house. A hot *kang*, tea and blankets, albeit they were covered with fleas, warmed me up. For an hour I told the story to half the village and jolly glad I was that I *could* tell it. Another hour in that infernal grave and I would have been frozen as hard as a cold storage pheasant.

CHAPTER XI

In the Desert at Last

AT the end of January, 1928, I cabled the American Museum to send out the Expedition's scientific staff. There were still certain difficulties about getting into Mongolia for bandits literally swarmed on the plateau. Most of them were concentrated in the area of Chinese cultivation from thirty to one hundred miles north of Kalgan. There the great caravan trails from Central Asia converge. Moreover, the bandits can find food and shelter from the villagers. No trade had been carried on with Mongolia for months. Not a caravan or motor car could move out of Kalgan without being robbed. Conditions became so bad that the merchants were being ruined and the brigands found that they had killed their source of revenue. As usual in China the matter was settled by the Chamber of Commerce. The local government officials agreed that certain "liaison bandits" would be allowed to enter Kalgan and make their own private arrangements with the caravan owners. Those who paid five dollars a camel could pass the brigand area unmolested. This form of protection is a common practice in China and works smoothly enough as a rule. Early in February thirteen thousand camels left Kalgan at one time, bound for Urga, Uliassutai, Kobdo and Hami in Chinese Turkestan. The motor cars which went to Urga paid the bandits one hundred dollars each for the hundred-mile passage through the danger zone—exactly a dollar a mile.

I started buying camels through my Mongols soon after Christmas, and by mid-February the one hundred and twenty-five for the Expedition's caravan had been assembled just beyond the brigand area. Meanwhile McKenzie Young and I had collected the supplies and equipment at Kalgan. Four thousand gallons of gasoline, one hundred gallons of oil, two tons of flour, a ton of sugar, thousands of pounds of rice and other food were packed and ready for the camels. There would be thirty-seven men, native and foreign, in the personnel. Thirty-seven men eat a great deal in six months and we could obtain nothing in Mongolia except game and an occasional sheep. Whatever we would use in the multiple activities of the Expedition must be provided in advance. There are no oases in the Gobi producing milk and honey, automobile parts or palæontological tools.

The boxes which carry our food and gasoline into the desert bring specimens out. Everything must do double duty. This rule extends even to the camels. Packing material for the delicate fossils would be unobtainable except for our caravan. A camel is an impossible sort of creature in every way. Nothing about him is like an ordinary respectable animal. Thus, when he sheds his winter's wool, it comes off in great strips and patches, making him look like an animated moth-eaten carpet.

We found that our camels shed at just about the normal rate at which we collected fossils, so when we needed packing material we simply pulled a yard or two off some camel whose wool was ready for plucking. But one must not be too careless about removing his overcoat, otherwise he will catch cold and die. In spite of his size he is a very delicate animal—or thinks he is. If he has a tiny cut in one of his great flat foot pads he will cry and moan like a baby. The Mongol's remedy for such an emergency is simple. The camel is thrown, his legs crossed and he is tied so tightly that he cannot move. Then a piece of leather is sewn over the cut, as one would patch a garment. It looks like a barbarous operation, but it is no more painful than shoeing is to a horse.

The camel's only food is the sagebrush and thorny Gobi vegetation. He wouldn't eat green grass if he had it. It would make him ill. The drier, the thornier and stiffer a plant is the better it agrees with him. The peculiarities of camels would fill a book.

Our caravan reached Kalgan early in March with five thousand other camels just down from Urga. Accompanying them was a bandit liaison officer. A few years before, he had been a respectable landlord of one of the motor inns on the Urga trail. I knew him well and knew that now he was a head brigand. What is more, he knew that I knew it. But it would have made him "lose face" to admit the fact. Therefore, we were introduced as though we had never seen each other. While he remained in Kalgan he posed as a general who could arrange protection for our caravan through his soldiers. Half an hour of tea drinking and extraneous conversation ensued before we got around to business. He suggested the customary fee of five dollars a camel. I offered one dollar. He knew that our boxes contained nothing that his brigands could use or dispose of and eventually we settled on one-half the usual rate.

Just before he left he brought up the matter of the one-hundred-dollar fee for each of our motor cars which would go out some weeks after the camels had left. I stalled on that. We were uncertain when we would leave —how many cars there would be—I would get in touch with him later! But I did mention that after all we hardly needed protection, for we would have thirty men on the cars, all would have rifles and, moreover, there would be a machine gun which could shoot two hundred bullets a minute. I did not make a definite statement. It would have sounded too much like a threat and wouldn't have been polite, knowing what he knew I knew. So I just sort of murmured it as though I were considering the matter aloud. But he got it. We parted with expressions of mutual esteem, and I dismissed the matter from my mind.

It was essential that the camels leave five or six weeks prior to the scientific party, for they acted as a supply base for our fleet of motor cars. Camels, when carrying the usual load of four hundred pounds each, can seldom average better than ten or fifteen miles a day, for they must stop frequently to feed. Our motors could do one hundred miles a day, so that we could move just ten times as fast as the caravan. When the camels left, I instructed Tserin, the Mongol leader, one of my most trusted men, to leave gasoline at a well near the trail and await us at a lámasery several hundred miles out in the desert.

On April 12th the staff with the exception of Captain Hill and myself left for Kalgan. Two days later, the

American Minister, Mr. J. V. A. MacMurray, his wife and sister, and Mr. Lewis Clark of the American Legation went up with Captain Hill and myself. The Minister and his party wished to see the Expedition start and to accompany us for a short distance beyond Kalgan. The Minister's presence was of much help to us in Kalgan. Not only did the officials hurry through the final passports, but we were relieved from paying the road tax which for our eight cars would have been a considerable amount.

When we left on the morning of April 16th the Kalgan authorities provided an escort of fifty cavalry to accompany us to the top of the pass. There was little danger for that distance, but they were taking no chances where an Envoy Extraordinary and Minister Plenipotentiary was concerned. After a few miles of driving sedately in the dust cloud behind the cavalry we assumed that the amenities had been observed. The horsemen were left to await the Minister's return, but at every corner of the road groups of soldiers popped out unexpectedly to present arms. This went on all the way to the village of Chang-peh-hsien, where the Minister's party left us. They got back to the foot of the pass earlier than the cavalry escort expected them and discovered the gallant soldiers in the act of robbing a caravan. The Minister started to take photos of the operations but the soldiers became very ugly and he had to stop. It was an excellent example of what happens all over China, for most Chinese soldiers are potential brigands. It is only a matter of opportunity. We spent the night at Chang-peh in a Chinese inn. Trouble might be expected on our next day's run through the bandit country and we wanted daylight for all of it.

A hard rain began early the next morning and continued all day, so that we could not start. I served out the rifles and ammunition for all of the men, and we planned exactly what to do in case of attack. The commander of the troops at Chang-peh told us that there were three hundred brigands in the region of Chep-sur, about sixty miles from Chang-peh, and that probably they would give us trouble. He also remarked that his furthest outposts were twenty miles from the village and that any armed man beyond that point was a bandit. Nevertheless, we did not worry at this rather grim information, because under ordinary circumstances twenty Chinese brigands to one foreigner is about the correct odds. The commander told us that a week before our arrival his troops had had a battle with about one thousand brigands. Since many of his soldiers did not return he surmised that they themselves had become robbers.

The next morning was bright, with a strong, cold wind. We left the inn at half past six and our cars made a most imposing spectacle. American flags and that of the New York Explorer's Club flew from every car and each man held a rifle in his hand. I felt certain that even three hundred bandits would hesitate before attacking us in the open.

For the first fifty miles we saw no armed men. The villages were almost deserted, the fields untilled, and few farmers appeared upon the road. Impressive indications of how the bandits had devastated the countryside! Not far from a small river called the "Black Water" we approached eight well-dressed and heavily-armed mounted soldiers. As they were much beyond the twenty-mile limit set by the Chang-peh commander, they must certainly be bandits and I told our men to keep them covered. They made no move to touch their weapons, but signaled that they wished us to stop. Keeping my revolver in hand I let them draw up alongside the car.

The spokesman said that they were part of a guard detailed by the Chang-peh commander to escort us through the brigand area. Of course, I knew that was a lie. Would we kindly stop in the next village to have tea and discuss the matter of our further protection? It so happened that the place where he had so cordially invited us to have tea was a fortified village where most of the robberies had taken place and about which we had been particularly warned. At my request for his credentials the "soldier" produced a letter written in Chinese purporting to be from the Chang-peh commander, saying that eight cars were to arrive with a party of Americans and to give them protection. But the letter had no "chop" or seal without which no communications are official in China. It would be like a letter of authority without a signature. The ruse was obvious. The bandits, being afraid to attack us in the open, hoped to entice us to stop in the village. While we were having tea we would have found ourselves covered by rifles and would have been either killed at once or held for ransom.

I told the brigands to leave at the double quick or we would shoot them where they stood. They wheeled their ponies and fled.

When we reached the village we found that it had been converted into a real fortress. Deep ditches on either side of the road prevented a car from turning off, and two diagonal, loopholed walls covered all approaches, Many men were ostentatiously lolling about the entrance as our motors roared through. They seemed considerably surprised when we did not stop.

Sometime later we learned the sequel to the story. A wounded bandit came to Hatt-in-sumu for treatment. He told Mr. Eriksson that the brigands were very angry because we had escaped their trap and were preparing a hot reception for us upon our return.

At our village Bato and the various families welcomed us with joy. Shackelford had brought many illustrated papers showing photos of these same people, taken the previous year. The women particularly were thrilled and handled the papers as though they were made of gold leaf. By the time we reached Hatt-in-sumu a violent gale was blowing and we welcomed the shelter of our tents.

In 1926 the mission station was moved from Hallong Oso, where it had been for some fifteen years. When it was established Chinese cultivation had not reached their region and the missionaries had Mongols living all about them. But as years went by the Chinese farmers pushed their fields farther and farther into the grasslands, driving the Mongols before them to new pasturage. Since their work was essentially among the Mongols, the missionaries had themselves to move in order to maintain contact with their people.

Mr. Eriksson had purchased an abandoned temple as the site of their new home. They lived in the temple until a new mud house had been erected. Now, there are seven buildings, including a dispensary and operation room, for

Mr. Eriksson's work is largely along medical lines. His medical training is sufficient to qualify him to handle ordinary diseases and accidents, and he has done an enormous amount of good among the Mongols.

The mission station nestles close to the base of low granite hills commanding a view of the rolling grasslands for many miles to the east. It is a charming spot and I cannot wonder that the Erikssons love it. Mr. Eriksson is an exceptional man. Alert and keen, able to turn his hand to almost any kind of work, he tells me that he is perfectly happy and hopes to live the remainder of his life and die in Mongolia. When he passes the Mongols will have lost a true friend. They appreciate little enough what he does for them. Eriksson is always ready to drive long distances in his little motor car to attend the poorest herder or one of the reigning princes. He even keeps on excellent terms with the lama doctors from neighboring temples and treats with sympathetic understanding the superstitions of his patients. The religious side of his work is difficult. After fifteen years the mission numbers only about forty converts, but the good he has done is measured in infinitely greater terms. There are two or three other mission stations, all Swedish, in the region.

Mr. Eriksson tells me that the most prevalent illnesses in the Mongols of his region are venereal diseases. He says that about ninety per cent of all men and women who have come under his observation have, or have had, syphilis or gonorrhea. This is due, of course, to the promiscuous habits of both sexes. Often, however, he finds healthy children from parents one or both of whom are afflicted with syphilis.

The lama doctors use a root called "tu-fil," which he thinks is a species of sarsaparilla, in treating syphilis. They give the patient very little to eat; no meat of any sort, and only a little soup. He is not allowed to leave the tent for several weeks. The Mongol name for syphilis is "the new time disease," but none venture to say how old it is or from whence it came. Doubtless syphilis is often contracted from pipes, but the lamas are by far the most potent agents in its widespread distribution.

Mr. Eriksson says that, after venereal disease, scabies is certainly the most common complaint. Because of their lack of cleanliness, almost every sort of skin disease that "flesh is heir to" appears at some time. Muscular rheumatism is almost universal due to exposure, cold *yurts,* et cetera. Eye troubles are very prevalent, such as conjunctivitis, cornea sores and ulcers. These are engendered by standstorms out-of-doors and constant smoke within the *yurt.*

Tuberculosis is decidedly uncommon. It is doubtless prevented by continual sunlight and the open-air life. Indigestion is a common complaint, and bronchitis in winter is almost universal.

Smallpox is prevented to a considerable extent by vaccination which is given to every child from four to ten years old by the lama doctors. They use the Turkish method. Crusts from a newly recovered patient are gathered and ground up with seed pearls and other substances which are supposed to weaken the germs. This powder is blown into the mucous membrane of the nose of the child. The patient is thus given a regular case of smallpox which may be light or severe. Often the children die. There are certain lama doctors who seem to have very few deaths to their discredit and naturally these are most popular. After vaccination the child is confined in the *yurt* and the parents try to prevent it from pulling off the crusts. Frequently an entire village will be vaccinated at one time.

Typhoid fever is often seen, but is not really prevalent, and typhus seldom. The lamas recognize infectious diseases and try to keep such patients segregated. Scurvy appears only in the spring, but at that time it is very common, because the Mongols then have no fresh meat or milk. They are living on meat killed the previous autumn and kept frozen during the winter. From the end of November to the beginning of June the Mongols kill no stock because the animals are all thin. Scurvy begins to develop in early April, but few people die of the disease. The natives are thoroughly familiar with its symptoms and know that almost any green vegetation will arrest its progress. Since nettles appear earlier than anything else, they eat quantities of this plant.

On Friday, April 20th, we left the mission station at six o'clock in the morning of a beautiful day. The air was so exhilarating and the sun so bright that we sang like school boys as the miles of plain vanished under the car wheels. The uninitiated would believe that summer had come to stay. But it was only a fortnight of perfect weather as a prelude to the blasting sandstorms and bitter cold of the next five weeks. Already the scanty vegetation had begun to show a faint trace of yellow-green. Bustards strutted like turkey cocks on the long hill slopes; bands of gazelles raced to cross our bows. Why they do I cannot explain. It is not only gazelles; wild asses, camels, ponies, sheep, cattle—every kind of animal on the plains seem to feel the same strange urge. They must pass in front of the car. Once over, some of them go on about their business; others cross and recross half a dozen times.

One gazelle gave us an exhibition of really high-class running. He was on a hard smooth plain when we came abreast of him about three hundred yards off to the right. He trotted parallel with us for a few moments and then broke into a run. For a furlong we held even at thirty-five miles an hour; then I stepped on the accelerator. He did likewise. Another push brought us up to forty miles an hour. He seemed rather surprised at that and slightly annoyed that anything should challenge him. After a quarter of a mile he evidently decided that the matter had passed a joke and he would end it right there. Then he really began to run. The flying legs were only a blur like the wings of an electric fan. His body seemed floating in space. I could not better forty-five miles an hour on that going and the gazelle drew in swiftly on a long slant, passing fifty yards ahead of us. Once across, he slowed down, gave a final leap as though he were on rubber tires, and stopped to gaze curiously at the car.

He had run a fairly good race, for he started three hundred yards away and we were going at forty-five miles an hour on a straight line. He was certainly doing sixty miles an hour. Can you imagine an animal not equipped with wings and having no gas tank, reaching such a speed even for half a mile? Of course a gazelle cannot maintain that pace very long. It has been developed only for that initial dash which takes them away from wolves that lie in wait behind rocks or in ravines. How far they can run I do not know, but ten miles I am sure of. Shackelford and I raced one on a great plain. At first he drew away from us and we could just keep his bobbing white rump patch in sight. Gradually we overhauled him and chugged steadily along at forty miles an hour, with the gazelle about sixty yards in front. His tongue was out, but he did not slacken his speed. The race was never finished, for after ten miles we got a puncture, but he didn't.

The trail was our former route of 1925 and as Major Roberts had already mapped it there was no delay for the topographic work. The tiffin hour found us fifty miles from Hatt-in-sumu, in a most picturesque spot among some giant boulders, and at six-thirty in the afternoon when the white escarpment of Shara Murun had begun to show as a purple line backed by the dull gold of sunset, we saw the camels of a great caravan grazing in the distance. Just beyond them a blue tent floated like a huge bird in the mirage. Then it settled definitely to earth and the American flag streamed out from its peak.

Our Mongols greeted us with joyous shouts. They had waited here, seven miles from the lamasery, because the feed was better. The camels were fit, the gasoline was not leaking overmuch, and generally all was well. We camped that night beside them. For the first time I felt that the Expedition was really under way after the two years' struggle with war and brigands, officials and diplomacy. The Gobi lay in front of us; our only opponents were the natural forces of the desert.

It was eight o'clock before we left camp the following morning. Until the sun was high we were comfortable in our fur coats and sweaters, shedding them in layers every hour. The Shara Murun River was dry and we drove up the broad valley on a trail like a race track. Gazelles were everywhere. We saw one herd in which both the desert and grassland species were feeding together—a very rare occurrence.

The Tukum temple lay white and dazzling four miles from Viper Camp, but we stopped only for a moment to greet the Mongols. By ten o'clock we had reached Viper Camp and swung north on the peneplain to Chimney Butte where, in 1925, a fossil quarry had been discovered. The old motor tracks were still visible as lighter green streaks where the snow and rain had settled in the depressions and given them a little more moisture than the surrounding plain.

Camp was pitched on a promontory close to the edge of the escarpment. Below us on three sides lay a gigantic relief map of painted badlands, gray and red buttes, ravines and canyons. In a narrow valley to the south of our projection were two *yurts* surrounded by a rampart of *argul* (dried dung)—a welcome sight, for it solved the fuel problem during our stay.

I knew it was unwise to be without wind protection in the spring but the top of the mesa was the only possible place for our fifteen tents; moreover there was no precedent for the utterly poisonous weather which was served out to us during the next month.

CHAPTER XII

Gobi Sandstorms

A BREATHLESS silence, suddenly dropping like a pall over the desert, brought me out from dinner in the mess tent the night we made camp. In the west a tawny cloud shot through with shafts of dull red boiled up out of the flaming pit into which the sun had disappeared. Already the purple line of distant mountains was blotted from the sky. A twisting, whirling skirmish line of tiny wind devils danced their way across the basin floor. Behind them the solid yellow mass advanced swiftly, ominously, engulfing the hills and canyons of the badlands like a devouring monster. Slowly we became conscious of a pulsing throb which beat upon our eardrums in an unearthly, soundless noise.

"Stand by for a blow! Rope the tents! Drive in the pegs! Get the cars around in front!"

At my shout the camp leaped into action. Hammers rang on the steel tent pegs; motors coughed and roared as the cars were lined in front to form a windbreak; there was a babel of shouts in Mongol, Chinese and English. The steady beat of the advancing gale changed to a rising hum, but as yet there was no wind. Suddenly, with a hiss, gusts of cold air like blasts from a snow field swept the camp. A moment later the storm had swallowed up the tents, tearing madly at the blue cloth. Clinging to the poles, we tried to keep our frail shelters upright. Blasts of gravel, like exploding shrapnel, brought blood to our faces. We could breathe only air thick with sand.

Granger and I weathered the first furious attack in the mess tent, but a great, jagged window had opened in the back. During a sudden lull I looked out upon a scene of chaos. The cooks were homeless, their tent only a mass of ragged streamers. Three of the others were down; the rest sagged drunkenly.

Thus began a night which I shall long remember. Perhaps one might not call it a real honest-to-goodness hardship, but for sheer discomfort it was one of the worst I have ever spent. Every one had his own troubles; so many of them, in fact, that there was little we could do to help one another. Most of the men elected to stay where they were under the flapping ruins of their tents, for those that remained standing were shelters in name only. The storm's first attack was the most vicious, but the others which followed in irregular salvos almost equaled it. Sleep was impossible. As for myself, I could hardly breathe. Seemingly a raging devil stood beside my head with buckets of sand, ready to dash them into my face the moment I came up for air out of the sleeping bag. There was something distinctly personal about the storm. It was not just a violent disturbance of the unthinking elements. It acted like a calculating evil beast. After each raging attack it would draw off for a few moments' rest. The air, hanging motionless, allowed the suspended sand to sift gently down into our smarting eyes. Then, with a sudden spring, the storm devil was on us again, clawing, striking, ripping, seeming to roar in fury that any of the tents still stood.

Thus it continued throughout a night of thrice the number of hours any respectable night should have. Now and then one of us would call to his companions in suffering, to be answered by smothered curses in three languages. The curses kept me cheerful. As long as a man can swear he is all right and in reasonably good spirits. After an indeterminable time we knew that morning had come by a change in the color of the enveloping cloud from black to yellow. That the wind would drop with the sun, we devoutly prayed. It did, and there were three hours of comparative quiet, allowing us to patch the least damaged tents. Then the particular devil that had been assigned to the day shift took up his duties with energy and vigor.

We spent a miserable twelve hours huddled in fur coats, for it had become bitterly cold. One could neither read nor write because of the falling sand; conversation was possible only through half-closed lips. Nothing to do but sit and think. Personally, I just sat.

At sunset there came a dead calm which lasted throughout the night, but the wind began again the following morning. For a month we had just one gale after another. The first one I have described; the others varied only in intensity. Day after day the men sat in the tents unable to work. When they could get into the badlands the results were gratifying. The first discovery was the course of an ancient stream which had run upon the surface fifty million years ago. The stream's bed could easily be traced where it was exposed in cross section—at the bottom the heaviest gravel; above it a finer layer, and still higher the lightest sand. Thousands of animals had died and fallen into the river during the Eocene, or Dawn, Period of the Age of Mammals. Parts of their skeletons had been buried and preserved in the fine sand and silt. In some spots they formed a breccia of bones, almost like a heap of jackstraws. Birds, fish, turtles, rodents, carnivores and other mammals lay in an indescribable tangle. It required the highest skill to separate one from the other. It is not spectacular work, for they can mean little to any but the specialist. Nevertheless, a collection of such varied types gives us a picture of the life of the world in those distant days. Each one is like a tiny flame illuminating an infinitesimal patch in the darkness of six million years.

As soon as the work was well under way I left the camp in charge of Walter Granger, Second in Command, and started for an exploration of the country to the southwest. With me were Capt. W. P. T. Hill, U.S.M.C., topographer, McKenzie Young, chief of motor transport, J. B. Shackelford, photographer, and Lieut. J. A. Perez, U.S.N., surgeon. Our object was to discover a trail which would take us into the far-western desert to Chinese Turkestan. This was the country we had intended to explore in 1926. Since that time Dr. Sven Hedin, the famous Tibetan explorer, had gone through it with a caravan of two hundred and eighty camels. No direct word from Doctor Hedin had been received since his departure two years earlier, but vague reports indicated that they had had a very hard time in a frightful desert; moreover, political difficulties had halted them for three months at the Turkestan frontier. Although Hedin's work was largely meteorological and quite different from ours, I did not want to follow in his footsteps. We were considerably handicapped in finding a new route by the fact that we had no permits for Outer Mongolia and could not risk a fight with the border guards if we crossed the frontier even for a short distance.

Our reconnaissance party left on April twenty-sixth in two cars, with food and gasoline for ten days. Tserin, the caravan leader, was with us in order that he might know where to take the camels if we found a practicable route. Running south and west we investigated every trail which appeared to lead in the right direction. All of them proved to be valueless, and we struck across the plains, traveling by compass. It was exactly like navigating a ship at sea. Captain Hill took our directions at every twist or turn and later plotted them upon his map. At night he determined our position by the stars. The going was not bad as a whole until we entered a wide range of low, granited ridges washed with sand. These gave us some hard work, but the worst was when we stopped for water at a well near a small pond. Apparently the gravel was bone dry. As Mac Young was driving away, his car suddenly sank by the stern and came to rest with the engine pointing at the clouds. Under the six-inch crust was a mass of jellylike mud. Experience has taught us that there is one thing to do in such cases, and only one. No use asking for a short cut to freedom. Collect stones, brush, bones,—anything solid—and sink a foundation beside each rear wheel. Every push of the jack presses the rock a little farther into the mud, but eventually it will hold. In this particular instance it seemed as though we would shove the stones straight through to New York. We can guarantee that at least one small part of Mongolia is firmly ballasted, for three of us collected rocks for four hours, and all of them went straight down. But inch by inch Mac built a foundation up to ground level and laid a neat causeway in front. Then, with the other car towing and all hands pushing, out it came.

A week of exploration served to show that the only possible trail to the westward was the one which Sven Hedin had followed. A Chinese with a caravan who had been gone two years told us that there was little water or feed; also that no motor car could possibly travel through the enormous sand dunes on the trail. But we have learned that native ideas of what one can or cannot do usually are valueless, and were not unduly depressed. For once they were right, however, as later events proved.

The evening of the last day before we returned to the main camp we came suddenly to the valley of the Shara Murun. Two twin buttes lay well out in the basin itself which at this point is very wide. The edge of the peneplain on which we had been traveling breaks off in an almost perpendicular descent to the branches of the river three hundred feet below. After considerable search we discovered a steep slope on which the cars could reach the valley bottom; and there our tents were pitched. The escarpment was beautifully exposed, but two hours' search gave us only a few fragments of fossil bone from both the highest and lowest levels. The deposit was evidently stream work and was largely composed of very coarse gravel. We could not hope to find important specimens, as everything would have been rolled and broken by the rapid stream action. Enormous quantities of gypsum were present in beautiful blocks and squares. Strangely enough, all the Mongolian deposits that contain gypsum are virtually unfossiliferous.

At nine o'clock it began to rain and continued steadily until that time the next morning. To climb up the steep slope, softened by rain, to the peneplain three hundred feet above us would have been impossible for less strongly built cars than our Dodges. The entire country was obscured by a dense fog, and we could see not more than a hundred yards in any direction. Captain Hill set a compass course for the Baron Sog temple, but long before we reached the trail a violent wind swept away the gray mist. Every hour the gale increased and when we reached camp at Chimney Butte we found all the men huddled in the tents. The last twenty miles had been a bitter fight against the bursts of sand and gravel which cut our faces and made it well-nigh impossible to drive. The windshield glasses of both cars were so badly sandblasted that they had to be removed.

After a consultation I found all the staff unanimous that we should still make the western trip even though it was necessary to follow in the main the route taken by Hedin's expedition. I decided to start the camels immediately so that they could get well on the way before we ourselves left Chimney Butte where the Expedition's staff all were busy and doing valuable work during the intervals between the sandstorms. From the caravan we removed food for three weeks and gasoline to carry the fleet five hundred miles. It was a most disagreeable day's work because the "false spring" had given place to cold sleet and fierce gales.

Granger, Pond, Spock and Horvath made an exploration trip to the east across the Shara Murun. The following day they returned having been driven back by the gale which blew incessantly. Nevertheless they had discovered several large exposures which warranted careful investigation. The camels had left and there was nothing that we could do. The terrific wind made work in the fossil pits impossible; it had become very cold; the drifting sand and dust so filled our eyes that it was difficult either to read or write; muffled in fur coats we could only sit and listen to the roar of the gale and the slating of the tents.

The wind had continued with only short interruptions for more than a week. Every one's nerves were on edge. During the intervals of comparative calm the men would work in the fossil fields for a few hours, but they could not endure the sharp particles of gravel which stung and cut their faces. The quarry yielded some excellent fossils. As it must certainly have been the slack side of a small river where there was virtually no current, the specimens were well preserved, although separated into individual bones. Even though we might have continued to work the stream bed with profit for some weeks longer, I decided to move camp to the new exposures to the east of the Shara Murun which Granger had discovered. We might find a more sheltered camp site and the mere fact of being in a new place with new vistas to look at would quiet the nerves of every one. Under such conditions I've always found it a good plan to move even if it is only a mile or two.

On May fifth we started happily for our new hunting grounds. As usual, Shackelford and I drove considerably in advance of the main fleet. Just after crossing the Shara Marun I wounded an antelope and left the car to finish it with my .38 caliber revolver. There was a safety catch in the holster to keep the gun from dropping out. In releasing it my finger slipped off the catch, pressed the trigger, and the double-action revolver exploded against my left leg.

The heavy bullet struck me such a terrific blow that I went down as though felled by a sledge. While Shackelford drove back in the car for the surgeon I discovered that the bullet had entered midway of the thigh on the left side, ranged downward and emerged below the knee, just nicking the distal end of the femur. After ascertaining that I could move my knee I lighted a cigarette and felt almost happy. Visions of a stiff leg and no more exploring, polo or hunting for the rest of my life had been distinctly depressing.

It was just such an event that we had always been expecting. During three years, with from thirty-five to forty men in the field, we had had no serious accidents. Something really was coming to us. Fortunately, we had a first-class surgeon. Through the courtesy of the Secretary of the Navy and Brigadier General Smedley Butler, commanding the Third Brigade of U. S. Marines at Tientsin, Dr. J. A. Perez, U. S. Navy, had been detailed to the Expedition. When he arrived he probed the long bullet course and put on a first-aid dressing; then I was carted to camp. The only view I had was the sky, but Granger said that we were in a shallow depression where the tents were somewhat protected from the wind.

In the afternoon Doctor Perez performed a very skillful operation with the assistance of McKenzie Young. The wound was filled with bits of leather and cloth. The surgeon had to cut along the bullet course for a considerable distance and clean it thoroughly and sew up the torn muscles. Shu, our mess boy, stood by to hand things, but tears streamed down his face so that he could hardly see. The doctor had given me such a shot of morphine that the world looked bright and rosy; in fact, I was rather pleased with myself. But the next day, when the morphine jag had worn off and a sandstorm raged, black clouds seemed to have obscured my particular sun.

I had a high fever and a fresh gale began to blow from the north. Blasts of sand and gravel swept into the tent in spite of everything that could be done. It was difficult to breathe. Because of the extraordinary precautions necessitated by the flying sand, the dressings were terrible. The tent was always filled with a yellow haze and the never-ceasing flap of cloth did not tend to soothe one's nerves. No greater compliment can be paid to Doctor Perez than that, in spite of such conditions, he kept the long bullet track absolutely free from infection. It began to heal perfectly and in ten days I was able to sit up during intervals of comparative calm.

Meanwhile work went on whenever the flying sand allowed our men to get out into the fossil fields. The deposits were comparatively rich, but the fauna was so like another veritable gold mine that I will tell of the strange beasts later.

On May 12th we had such a terrific sandstorm that no one could leave the tents. In addition the temperature dropped many degrees and the wind was bitterly cold. Granger, Horvath, Hill and Shackelford were to make an exploration trip east as far as Iren Dabasu (Erhlien), and they left the next day even though the wind was still blowing a full gale. A bright sun raised the temperature somewhat but still they needed their heavy fur coats. But the following day was so cold that ice formed in the water buckets even in our tents, and the sandstorm continued with unabated violence. I think that no one who has not endured sandstorms can understand how difficult it is. Physically weak, in continual pain and with fever, it became well-nigh unendurable to me. Often I had to bury my head in the blankets to keep from screaming. It seemed that something in my brain would crack unless there could be a rest from the smash and roar of the wind, the slatting of tents and the smothering blasts of gravel. But not a respite did we get. Sometimes there was an hour of comparative calm as the wind died with the sun, but before the after-glow had faded from the sky a new gale had burst upon us.

At last, on May 16th, after a night of bitter cold, the sun lay soft and warm in the tent. Only a gentle breeze played over the green-tinged plains. I felt a surge of new strength and a great desire to be up and out. Mac Young had improvised a useful crutch out of sticks from Shackelford's portable dark room and I could hobble about quite comfortably. At a spring three or four miles away, the sand grouse came in hundreds and Young, the doctor and I drove there in the open car. Placed behind a high mound with a patch of water in front I had excellent shooting at the birds which came in like bullets to circle about the stream. It was a veritable release from prison and I felt new life soaking into me from the sun-drenched earth.

About seven o'clock of the next evening Granger and his party returned from the Iren Dabasu trip. It had been very successful. They discovered that the sedimentary basin continues for more than a hundred miles to the northeast and is well exposed in many places. Excellent prospects were offered for palæontological investigation. While they were at Iren Dabasu Granger had spent a few hours at the Cretaceous dinosaur beds where Johnson and Kaisen had worked for several weeks in 1923. He found a quantity of dinosaur egg shells and was convinced that the exposure was an extensive nest site for the duck-bill and other dinosaurs.

They also obtained unsatisfactory war news at the telegraph station. The so-called Nationalists had taken Techow, only ninety miles from Tientsin, and were steadily pushing northward. There had been serious trouble in Tsinanfu, Shantung, which had been quelled by the Japanese with a loss of some two thousand Chinese soldiers. Colonel Holcomb, Commander of the U. S. Legation Guard, reported Peking quiet. Only a few cars had ventured on the Urga road because of the bandits.

The next day we were visited by an official of Durbet Wang, the prince of the district in which we were working. He had with him six heavily armed and picturesque

Mongol soldiers. Subsequently we discovered that more than twenty others were concealed behind the surrounding hills within easy shooting distance of the tents. They were taking no chances of riding into an unfriendly camp. The official and one of the others spoke Chinese well. I received them in the mess tent, giving them tea and cigarettes. They asked about our work and we showed them the fossils and explained that these beasts lived in Mongolia long, long ago. They seemed really interested and impressed. After photographing and giving them a Victrola concert they rode away in good spirits.

On May 19th I drove over to Baron Sog lamasery to ask if we could leave our fossils there until our return. They had refused in 1925 because of the "rinderpest," which the lamas had said was the result of our digging up fossil bones. The spirits were angry. After considerable talk with the head lama and a promise of fifty dollars, he agreed that the spirits might be pacified if we kept it a profound secret among ourselves.

CHAPTER XIII

Westward Into the Desert

MY strength had returned so rapidly during four or five days of fine weather that the doctor agreed to let us start westward and we left on May 20th.

Not more than five miles from our camp that night was the residence of Pailing Wang, prince of the district. The name means "Prince of Larks" and the entire Pailing Miao region is known as the "Land of the Larks." The Mongolian lark, which is the most abundant bird both in the grasslands and on the desert, is highly prized by the Chinese. They are excellent singers and mock the songs of other birds and the miaow of a cat in a remarkable way. In this region as elsewhere in the grasslands, Chinese come to catch the nestlings for market. In Peking a good singer brings five dollars (Mex.) and specially trained birds much more than that.

At Pailing Miao, the "Temple of the Larks," we were told that our caravan was awaiting us fifteen or twenty miles to the west. Why they had stopped I could not imagine. The trail led us up and down a series of grassy hills, broken by many outcrops of hard rocks. Spock was happy, for the structure changed continually and he obtained most interesting data on the old rock formations.

Only thirteen miles from Pailing Miao we discovered our caravan camped near the trail. Tserin said that there was a Chinese *yamen* (official dwelling) a few miles away which had been giving all caravans so much trouble that he did not dare go on without our presence. The feed had been so poor all the way from Shara Murun southward that the camels were very hungry. Several had died, and I decided to send back thirty-one of the weakest animals which we could relieve of their gasoline loads. The lamas at Baron Sog Monastery would care for them and they would be fat and strong upon our return.

The *yamen* proved to be a tax station between Chahar and Suiyuan. The officials were decidedly nervous about asking us for taxes, but as it was a legitimate provincial station and their credentials were quite in order, I was willing to pay ninety-one cents per camel, as did the regular caravans.

That night, just before dark, we heard joyful shouts in Mongol and Chinese. Over the hill, out of the sunset glow, came a small caravan; the men were running ahead wildly embracing our Mongols. Soon we saw that it was a detachment of Hedin's expedition returning from Turkestan. There were twenty Mongols and three Chinese. Most of the former had been with us at some time and were old friends. Sitting in a circle there in the softly glowing twilight, with the aid of rude maps traced in the sand, we heard their story. Hedin had had a good deal of political trouble at the Turkestan frontier, but finally had been allowed to proceed with some of his expedition. These men were returning with the collections to their homes north of Kalgan after more than a year's absence.

All of them were gaunt, with hollow cheeks and

hungry eyes. For weeks their only food had been camel meat and a few split peas. Money they had in plenty, but there were no Mongols who could sell them sheep. Little water, no food and few inhabitants; sand dunes of vast extent through which they had wandered for days seeking a passage for their dying camels. A country of utter desolation. It was not a cheerful picture. Nevertheless, I decided to push on as far as possible, for there might be fossil deposits which would be worth the effort. I hoped, too, that we could circle the sand to the north and still keep in Inner Mongolia.

The topographers went ahead the next day, but I remained with the others to see the caravan safely past the *yamen*. That night a caravan of Shansi men with one hundred and fifty camels camped near us. They were carrying goat skins, on the return trip from Urunichi to Kwei-hua-cheng, and had been ten weeks on the road. With them were two Chinese passengers; they rode in a felt-lined box on either side of the same camel. The boxes looked exactly like dog kennels. Although they were well padded inside, it must have been torture to ride in them because of their small size. As soon as the camels arrived, the caravan men pitched a small tent for the two Chinese passengers, who enthroned themselves upon a pile of bedding and gazed out upon the world in a most superior fashion. Having spent the day in their kennels one would have thought that they would like to stretch their legs a bit, but their only exercise consisted in walking from the camel to their tent.

The Gobi caravan men are hard-bitten fellows, with a certain reckless swing about their carriage and a terseness in conversation which suggests the wild, free plains on which they spend their lives. They invited us to have tea with them. Fortunately, I can understand the Shansi dialect and was able to get a good deal of information regarding the country to the west. They told us that the great sand began near a temple called Shandan Miao and that it would be absolutely impossible to cross it with cars. Still I was determined to push on and see for ourselves.

We left the *yamen* at eight o'clock the next morning and soon met the topographers. They had camped at dark the night before near a well in the stream bed. Hardly had their tents been pitched before fifteen villainous looking Chinese armed with Mauser pistols rode in and camped near them. They were opium smugglers from Kansu who had sold their drugs in Kwei-hwa-cheng and now were returning for another consignment. We had heard of their activities. They took a sheep whenever they needed it for food, but confined their robbing to fast ponies except when some especially rich prize came their way.

Granger arranged the four cars in a square with the tents inside and the spot lights illumined all of the surroundings. Two men remained on guard all night. Although the bandits doubtless would have greatly liked our rifles and cars, they were afraid to attack as they saw that our party were well armed and watchful. After the smugglers made camp, one of them rode back along the trail to see if they were being followed. It was lucky that the topographers had four cars; if there had been but one or two they would certainly have had a fight.

These opium runners are usually much more courageous men than the ordinary Chinese bandits. On the outward trip, loaded with opium, they would not stop for ordinary robbery. They want only to be let alone. But on the return trip they become much more dangerous individuals. Motor cars would have been a heaven-sent method of rapid travel. As a matter of fact, when the Expeditions ended in 1930 I sold two of our cars to Chinese who, I later heard, used them for running opium from Kansu.

At daylight next morning the bandits left. Our detachment passed them farther to the west. They were traveling at a fast trot which would average about six miles an hour. Some Chinese living beside the trail told us all about the smugglers, who were well known.

In the afternoon we arrived at a deserted monastery, Shirigi-in-sumu, which was the farthest point we had reached during our reconnaissance in 1925 and where I decided to await our caravan. In the short run of forty-two miles from the *yamen,* we had come into an entirely different region. The grasslands had gradually given place to an exceedingly arid desert, where the only vegetation was sparse camel sage, scattered over rolling sandy hills.

The deserted temple stands in a great amphitheater among the hard rock ridges, and is approached through a narrow gateway in the southernmost range. It is a desolate spot, for the black lava ridges and yellow sand are relieved by hardly a touch of green. We were told that there were wild sheep in the mountains and I shot both species of antelope in the lowlands to the south.

Chukkar, or red-legged partridge and quail were fairly plentiful on the ridges.

The next day, May 27th, was quite the hottest of the spring thus far. There was a dead calm until about ten o'clock, when strong whirlwinds began to appear everywhere on the desert. McKenzie Young went back to the caravan to bring up a load of gasoline. He returned at four o'clock in the afternoon and reported that the camels were coming on very slowly because they were weak from lack of food. We had already discovered that on this trail the caravan men carry dried peas for their camels. In no other part of Mongolia had I traveled a region which was so desolate that there was not sufficient vegetation to feed a camel.

May 29th was an exceedingly hot day with a strong wind. It died at sunset, but about ten-thirty in the evening it began to blow hard. By eleven o'clock one of the worst gales I have ever experienced in Mongolia was howling down through the rocky gateway from the north. We lashed and weighted the tents, but it was impossible to keep out the blasts of sand and gravel.

The next day we spent a miserable twelve hours huddled in fur coats, for it had become bitterly cold. The wind dropped a little late in the afternoon and the caravan arrived. I took out supplies and directed Tserin to wait at the river until I sent him word. The camels were becoming so weak from lack of food that it was obvious they could not go on under such conditions.

Much to our disappointment, we found that the topographers had made little progress. They had been stopped by Mongol soldiers about ten miles from the ruined temple. The Mongols had posted men with rifles on several hilltops evidently expecting a fight from our people. Granger had explained our work to them, but they would not allow the topographers to go forward until they had reported to their prince who was some fifty miles away. Since the Mongols had not returned, I told the topographers to break camp and carry on the work. Granger said that his party had fared somewhat better in the sandstorm than we had, for they were under the lee of a hill, but that two of their tents had been badly ripped.

The entire region was more arid and depressing than any that I have seen in Mongolia. It was totally unlike the Gobi as we had seen it north of the Altai Mountains. Each day the going became heavier and we were literally fighting for every mile. There were none of the wide spaces of the northern Gobi. To be shut in, oppressed, by naked hard rock ridges without the majesty of size; to plow through drifting yellow sand in valleys neither wide nor narrow; to look upon the bleaching bones of camels, dead from thirst or hunger—that is worth enduring only when it is producing valuable results. Thus far we had obtained almost nothing and prospects for better success in the future were very dark.

On June 2nd we reached a fine rolling gravel desert much like the peneplains of the north and camped at a great exposure of red sediments. The spot gave every indication of splendid fossil-bearing deposits, and we thought that at last we had found what we had so long looked for. In a dry stream bed in the center of the basin Shackelford discovered a moist spot. He dug a well, lined it with a gasoline box, and in a short time we had an abundant supply of clear cold water.

The next day dawned bright and warm. Every one prospected the badlands, but Buckshot was the only man who found bone. His discovery proved to be a large Sauropod dinosaur resembling *Diplodocus,* a species sixty or seventy feet in length. The bone was in very bad condition and Granger soon saw that the best he could do was to take a few distinctive parts which would serve as the basis for a description of the animal. Doubtless it is a new record for Asia and pretty definitely placed the formation as Lower Cretaceous near the end of the Age of Reptiles. It was the only fossil of value which we had obtained since leaving Shara Murun. Strangely enough, the most diligent search of the acres of beautiful exposures did not yield another scrap of bone.

I decided to make a reconnaissance to the west with two cars, leaving the others to await my report. The terrain was rapidly becoming impassable for our cars because of the drifting sand, and we had the statements of Hedin's party and the caravan men that motors could not possibly pass the great sand area farther west. We were exhausting our gasoline and men, wrecking the cars, losing many camels by starvation, and obtaining virtually no results. Unless there was a very decided change for the better I determined to abandon the western exploration.

That night over the radio we got time signals perfectly from Cavite, and a few moments later Shackelford picked up our call letter from the American Legation in Peking. The message came in so distinctly that even in my tent several feet away I could hear the dots and dashes.

Shackelford, Horvath and Hill took it down independently, then Granger and Shackelford squabbled like two crows over the translation, for the message was in code. Meanwhile the rest of us impatiently awaited the reading. At last we had the following:

> "Marshal Chang Tso-lin retreated yesterday in Manchuria. The Southerners are expected in Peking very soon. Everything normal in Peking and no trouble is expected. All are well. It is rumored you are returning next week."

It was an interesting message for us. Since our permits were all from the Chang Tso-lin government, they would be worthless with the southerners. It was fortunate that the change had taken place so early in the spring, for by the time we were ready to return in the autumn an adjustment probably would have been made. Just how the "rumors" of our return could have arisen we were unable to understand. Still I know of no other city in the world where there are so many rumors as in Peking.

Our Chinese, all northerners, were greatly excited by the news. The invasion of the southerners was that of an alien people for them. To add to their depression, there was an eclipse of the moon that night. We watched the black shadow slowly swallow the bright orb until only the lower edge remained unobscured. The Chinese were sure that it augured badly for the future of their country!

The next morning, Young, Hill, Spock, Perez and I left with two cars for the west. A few miles took us off the peneplain and into a lava ridge where the going was extremely rough. The black rocks were drifted with yellow sand and showed hardly a trace of green. Descending to a dry river bed, we went directly southward between high barren and sand-swept hills. It was inexpressibly desolate; like being enclosed in a sand-walled prison.

Pushing the cars every few hundred yards, we emerged at last through a narrow gateway into a great open basin. It was the edge of the vast depression which we had tried unsuccessfully to reach from the north in 1925 and where it appeared probable there would be fossil-bearing rocks.

In the distance a beautiful red mesa and many isolated buttes showed encouragingly against the sky. We thought that at last our efforts had been rewarded and that we had reached the Promised Land. But a closer view revealed that the escarpments were banked almost to the top with blown sand. Every one of the buttes in sight were in the same discouraging condition.

The floor of the basin showed as a yellow blanket, but we pushed on southward. I use the word "push" advisedly, for that exactly expresses our progress. The cars could go only a short distance under their own power; then every one out, and with rope-bound canvas strips laid down in front, strain and push until the car had reached an area of harder sand. Such work is difficult at any time, but under a blazing sun with the temperature at 145° F. it becomes doubly exhausting.

Two *yurts* were pitched well out in the basin and occupied by Chinese who appeared to be as desolate and miserable as the country in which they were living. They told us that the name of the region meant "The Very Bad Place."

Three abandoned camels were lying beside the trail, alive but too weak to raise their heads. Gobi caravan men never kill an exhausted camel while on the march. They believe that its spirit might follow the caravan and cause trouble among the other camels. Instead they leave it to die a lingering death of thirst or starvation. After these three were dead, the resident Chinese would strip them of their skins, which can be sold for five or six dollars.

Eventually we reached a flat ridge near the end of the red mesa. It was only a quarter of a mile wide and we looked down into another basin spotted with tamarisk bushes about two feet high growing out of a heavy sand blanket. Another ridge separated this basin from a third filled with deeper sand and smaller tamarisks.

At last that was crossed. A fourth and worse basin shimmered with waves of heat. In its center the sand defeated us. Utterly exhausted from pushing, we left the cars where they were and dragged ourselves forward half a mile to the western rim. A depression so wide that its limits could not be seen lay before us like a yellow blanket, specked with olive green. It seemed to stretch on to the edge of the world, a lifeless sea of burning sand. To enter it with motors was absurd; only camels could carry on to the other side. It marked the end of the trail for us.

The return trip was fairly rapid because our outgoing tracks gave much assistance. When we reached camp I told the men that the western trip must be abandoned. Motor cars simply could not be used satisfactorily in this region. Conditions were utterly different from those in Outer Mongolia where we had worked in previous years.

There the desert is of steppe formation and has a gravel floor with sand only in infrequent dunes, long and narrow. Here, we were on the northern edge of the Alashan Desert from which the drifting sand sweeps up to meet the true Gobi. Camels are the only possible transport for this region, and even they have great difficulty because there is so little feed. Had we been obtaining important results in the different branches of science represented in the Expedition, it might have been worth while fighting the sand or leaving the cars and continuing on camels. But the returns for our investment of work and time were virtually nothing.

Abandoning the western trip meant revising the entire plans of the Expedition. A great area of Eastern Mongolia still remained almost unknown. There we would go as fast as we could extricate ourselves from the devouring sand.

CHAPTER XIV

Eastward Into a New Land

RETURNING, we found Tserin encamped at the river a few miles from the ruined temple. Five of the camels were dead; many others could barely drag themselves along without loads. This trail to Chinese Turkestan must take a terrible toll of both animal and human life. As far as we had gone the way was marked with dead and dying camels. Nowhere north of the Altais had we encountered a region of such utter desolation, and all reports said it was much worse farther west.

About twenty-five miles farther on we camped, near the base of the lava mountains, in order to give Doctor Perez and Horvath a chance to hunt big horn sheep. Neither had ever seen a live *argali* and they were both wildly excited.

We could find no ponies or Mongols to guide the doctor, but at daylight he and Horvath went out alone. It was a blistering hot day, and at noon Horvath returned, completely exhausted. They had found sheep in the first hour, missed a big ram, and he had killed a small one. The doctor continued alone.

At five o'clock we saw a weird figure slowly approaching camp. It looked like a man, but had strange projections in the region of the ears. As it came nearer we made it out to be our surgeon. He was dragging painfully, just able to walk. A huge pair of horns rested on his head while the neck-skin flapped disconsolately about his face. Soon he shifted it to his shoulders. Then he tried both hips. When he saw that we had discovered him he threw his burden on the ground, kicked it disgustedly, and collapsed upon a rock. Young drove out with a car to collect the remains.

The doctor had had a hard day. He killed his sheep about ten o'clock many miles from camp. Then he discovered that he had forgotten to take a rope. A sixty-five-pound sheep's head is about the most awkward thing to carry that I know of, if one cannot make a sling to get it on his back. There was not a vine or a green branch on those bare hills and the sun beat down with one hundred and thirty degrees of heat. Perez had his field glasses, canteen, knife, rifle and ammunition, in addition to the head. By the time he had walked five miles and climbed up and down four ridges, he had come to hate the thing. He was sore all over and rubbed raw by the horn ridges. But it was a magnificent specimen and his first *argali*. The horns measured forty-one inches in length and nineteen inches in circumference at the base. Any sportsman can understand why he stuck by it. He said that he knew he never would have returned there if he had left it and gone back to camp. The day was too hot, the way too long and the mountains too rough. He swore that his sheep hunting had ended forever, but the next morning I noticed that he threw out tentative suggestions about another day's delay.

The doctor had seen two other fine rams in his few hours' hunt, and we examined a dozen pairs of enormous horns along the base of the ridge. The Mongols assured us that the low mountains were full of sheep. Possibly no other white men have hunted there and the natives have no rifles. Leslie Simpson, the African sportsman, now holds the record for this species of sheep, beating one of mine by an inch or two in length.

At the *yamen* there was a new lot of soldiers and officials. These were men of General Yen Hsi-shan, who had already replaced the Chang Tso-lin adherents. Although the former soldiers had collected taxes for one year in advance from the poor settlers, the new officials were busy levying their own particular taxes. The Chinese farmers thus are harried by every change of government in their particular region. How they manage to exist at all is a mystery. If they refuse to pay the taxes they are often killed or beaten into submission.

The return trip to Shara Murun was accomplished without incident. We stopped north of Hospital Camp, on the edge of a great escarpment. The tents faced east, and a beautiful plain swelling to rounded hills lay to the south. Northward, behind us, was the saw-toothed rim of the bluff from which we looked into a great basin splotched with streaks of bright red sediments. The exposures along the escarpment are mostly gray.

Just below the tents water from the spring formed an intermittent stream in the basin where sand grouse continually came to drink. Two ruddy sheldrakes chattered and called, and twice we saw a fox disappear over the basin's rim; he must have had a wife and family in a burrow of the escarpment.

Captain Hill and Dr. Spock immediately began a

topographic and geologic map of the badland area. Pond hunted artifacts, finding several stations close to the spring, and the palæontologists all discovered skulls or jaws of titanotheres and *Hyænodonts* before the first day had ended.

It was necessary for Young and Horvath to return to the caravan with two cars, to bring gasoline and food to us. I estimated that because of the weakened camels the caravan could not arrive in less than three weeks. In the meantime I wished to make a reconnaissance of the region east of the Kalgan-Urga trail. Even the inaccurate maps of Mongolia left it vacant. Its topography was unknown; not a ray of light penetrated the darkness of its geology and palæontology except just north of the Chinese frontier. Without additional food and gasoline the exploration would be impossible. Relaying supplies by car in this way was most expensive, but there was no other method due to the breakdown of our camel transport.

The work went on most satisfactorily. Granger discovered half a dozen skulls and jaws of small carnivores, and several beautiful skulls and jaws of titanotheres were found by various members of the party.

One day Pond made a most important discovery. A picturesque Mongol rode into camp carrying an antiquated flintlock gun. Pond noticed that its striking-flint was a beautiful Dune Dweller scraper. The old hunter said that he got it twenty miles away where there were many of them. Next morning Pond collected the Mongol at his *yurt*. It required considerable patience to entice him into the car and once there to prevent him from leaping out when the motor started. But his fright soon changed to exhilaration. He took Pond on a wild cross-country ride, for he evidently thought that a car could go wherever a horse could travel. Pond said that the car did everything but fly.

Their destination proved to be a most unimpressive-looking spot, not far from the great lake basin southwest of the Baron Sog lamasery. It was merely a shallow basin among dead sand dunes. But the artifacts were there in thousands and Pond returned to camp tremendously elated. For his work the Mongol received a silver dollar, three gas tins, a bottle, a package of cigarettes and a jam tin. He promised to guide Pond to two other localities, but his sudden wealth excited the jealousy of a wandering lama who happened to be in camp. The wretched priest told our Mongol that it was "bad joss" to show such places to foreigners and forbade him to go again. It was obvious that Pond would have to camp at the station and spend a week or more of careful study. In the meantime he found another great station south of Hospital Camp.

Young and Horvath returned on June 14th after having made a record trip to and from the caravan. With gas and food in hand I decided to make the eastern reconnaissance at once.

To the north of us Granger had discovered a great exposure of sediments which obviously were identical in formation with those we were on. I asked him to prospect them more carefully and, if he deemed it advisable, to move camp to the new locality while we were in the east. When he did so, Pond could go alone to the Dune Dweller site and finish his work.

On June 16th I started east with two cars, and Young, Hill, Spock and Perez, also three Chinese and a Mongol. We had gas and food for a week, and expected to drive seven or eight hundred miles. On the first day, before reaching Erhlien, the pinion gears on one of the cars broke. This is an accident which never had happened before with us, and was due to the terrific pounding which the gears had sustained in the sand on our western trip. Fortunately we had another pinion with us and we camped while Young changed it. The break had occurred in a sand wash, near a well, but in a difficult place to work. It was finished in time for us to reach Erhlien at four-thirty the next afternoon. The telegraph operator was a genial old Chinese whom we knew from former trips. He said that no cars had passed on the Kalgan-Urga trail for a month, but that he expected four to arrive that evening. General Yen Hsi-shan was then in Peking and the political situation was quiet. We camped at our old historic site where we had discovered the first strata of the Age of Reptiles and the first dinosaurs in 1922.

That evening the four cars arrived. One, a five-passenger Dodge touring car, was carrying sixteen hundred pounds of baggage and twelve passengers. The auto was almost obscured by human bodies projecting at all angles. The chauffeur, a Chinese, reported many brigands near Cheptsur.

After leaving ten cases of gasoline at Erhlien for our return journey, the next morning we followed a trail eastward which eventually turned to the south. We zigzagged back and forth for many hours trying to find an east and west trail. Captain Hill took compass directions at every turn, and our navigation was carried on exactly as if we were on a ship at sea. Late in the afternoon we found a well where we camped. We were then about fifty miles east and a little south of Erhlien, although we had traveled one hundred and twenty-five miles. The well lay in a deep basin surrounded by poorly exposed bluffs. At the nearest point we discovered fragmentary fossils and great masses of fresh-water clam shells arranged in strata. Evidently we were in an ancient lake basin and the indications were that it might be Pliocene. The late Tertiary formations were what we were hoping to find in the east as they are notably absent in the western Gobi.

Next morning while the cars were being packed Doctor Perez and I had good sand grouse shooting at a deep pit half filled with rain water, where the birds came to drink. Then we drove over to an exposure five miles from camp and just off the trail. It marked a part of the northern rim of the lake basin. The exposures were bad, consisting mostly of coarse rubble, but on the very summit of the escarpment Buckshot discovered a mastodon skeleton partly exposed, also rhinoceros bones. This appeared to identify the formation as Pliocene and we were correspondingly elated.

Continuing eastward across country, we entered upon the largest and flattest peneplain that I have seen in Mongolia. It was covered with sparse grass, but there was not a sign of water or evidences of any except old camp sites. Evidently the Mongols come there only in winter when snow can be melted for water.

On a low hill was an *obo* (religious monument) visible for a long distance. It is used as a landmark for those

Mongols who venture out upon this vast waterless plain. It is a "lighthouse" in the desert.

Some miles beyond the *obo* the plain gradually began to slope downward and in a valley we discovered two *yurts* and a few sheep. The Mongol told us that there was a well a mile away—the only water for a long distance. After taking our position, Captain Hill reported that we were only a few miles south of the Outer Mongolian border. Of course we could not cross, so next morning we started southward and eventually found a trail that was fairly good but led us through a succession of hard rock hills. For tiffin we halted three miles beyond a well where there were three *yurts* and some very friendly Mongols. When I started the car after the meal there was the sound of a suspicious grating in the region of the differential. In a few moments it became evident that the bevel, or ring, gear had broken. It was a bad accident for we had no spare with us although there were several in the main camp.

Young immediately removed the rear end and found our fears realized. There was nothing to do but tow the car back to the well and leave it with two of the Chinese while we returned for the spare part. This accident, like the broken pinion, was due to the terrible strain of work in the sand on the western trip. In all our use of Dodge Brothers cars, in the previous years of exploration, nothing had ever happened to the driving gears and we had come to think of them as unbreakable.

The next morning we left at eight o'clock. We had given almost all our remaining food to the two Chinese, and money with which they could buy sheep from the

Mongols. Seven men and gasoline made a heavy load for the single car, and we had many doubts about being able to negotiate the rough country between us and Erhlien. There was nothing else to do, however, and we started off full of hope if not confidence. Captain Hill laid a compass course for Erhlien, which we estimated to be about one hundred and twenty-five miles in a direct line. We saw only two or three Mongols and they knew little about the country or trails except in their immediate vicinity. It was a trying experience to negotiate the great waterless plain. Had the overloaded car broken we should have been in a serious situation. With little food and water and no Mongols, it would have meant a long foot march and I doubt if all of us would have survived. However, the car plowed steadily onward, bumping over the rough terrain, without a pause. Eventually we saw what we supposed was our "lighthouse" *obo* on the plain, but it did not check with Captain Hill's observations. It was much too far to the south. But Hill was certain that he could not be mistaken, and we decided after a consultation to continue on his course. Subsequently we discovered that there were two *obos,* almost exactly alike.

At eight o'clock in the evening we reached Erhlien pretty well exhausted from the rough ride and nervous uncertainty. It was evident that we had been traveling in the most extensive basin yet discovered in Mongolia and that it was probably formed mostly of late Tertiary sediments. It would yield excellent opportunities for future work and doubtless give a new fauna. At the telegragh station welcome messages were awaiting us from Peking.

One from the American Minister, Mr. J. V. A. MacMurray, read as follows:

> "Northerners withdrew toward Manchuria fortnight ago. Peking government discontinued. Peking-Tientsin area and northwest peacefully occupied by Shansi Nationalists. Advance of Shansi and Kuomintang toward Shanhaikuan still progressing. Good luck. MacMurray."

We heard from the telegraph operator that when the northerners withdrew from Kalgan and before the Nationalist troops had occupied the town, a thousand brigands had swept down upon it from the plateau. They had looted certain sections of the city and demanded a ransom of $500,000 from the Kalgan Chamber of Commerce. They remained in possession for two days and left as the advance guard of the Nationalist soldiers appeared.

We felt sure that Granger would have shifted operations to Urtyn Obo, the great exposures which he had discovered, fourteen miles north of Spring Camp. Therefore we headed directly toward that spot. On the way I shot a gazelle at long range. It proved to be a female carrying young. While I was lamenting the fact, Dr. Perez jumped out of the car and performed a neat Cæsarean operation. Lifting out the young antelope which was just ready for birth, he employed artificial respiration and soon had it breathing naturally. It would doubtless have lived had it been uninjured, but my bullet had shattered both hind legs.

At four o'clock in the afternoon we saw the blue tents

of our camp floating in a beautiful mirage. It was pitched on the rim of a great badland basin. We were very tired and very, very dirty. After tea Granger led me to a projecting buttress behind the tent. We looked out over a wild chaos of ravines and canyons and gigantic chasms, yellow, red and gray. A huge *obo* built by the Mongols as an offering to the gods of this fantastic spot crowned a sentinel butte. Sunset shadows filled the mysterious chasms with soft purple masses. Pinnacles and spires stood in silhouette against the sky. Over this tumultuous land sea lay the exquisite calm of a desert evening.

Quietly my old friend told me of the first week's finds. His steady blue eyes filled with the light of affection and happiness as he saw me thrill with excitement.

"There," said he, pointing to the topmost layer of golden sands, "is where we find the bones of *Baluchitherium*. We do it with field glasses. It is a new kind of fossil hunting. I've never done it before. But it is the best way. We walk along the ridges and look across the ravines. See that white spot over there? That is a cervical vertebra I discovered today from this very spot. You will like this kind of prospecting.

"Below the yellow sands," Granger continued, "in that thick mass of red, there is nothing. So far we have found only a few bits of turtle shell in its whole extent. That gray-white stuff under the red is full of titanotheres. Every one of the boys is working on a skull and they are all good. The far layer of pink contains some fine things, and way out there in the bottom of the basin in those white sediments we seem to have a new fauna. Probably it is merely a facies of the upper gray, but anyway it is different. Thomson found an extraordinary rhinoceros there—a little fellow. I've never seen anything like it."

So the story went on; a story which set my heart to beating happily. After the long weeks of discouragement, the days and nights of pain and heat and utter exhaustion, at last the desert had paid its debt. I slept that night with a great load lifted from my spirit.

CHAPTER XV

Discovery of the Baluchitherium

IN the morning I tried the new method of field-glass prospecting. It was fascinating. Walking out on the narrow ridges between the ravines I could look across forty or fifty yards to the other side.

The powerful glasses magnified the surface of the yellow sands bringing out in stereoscopic relief every pebble. Success was immediate. In the bottom of the first chasm I saw the white fragments of a huge bone. Following it up with my eyes, I found the source. A mass of bone lay just under the canyon's rim. Ten minutes' walk brought me to the spot. Most of the fossil remains were useless; simply a jumbled heap of great fragments without form or character. Evidently a limb bone broken by the action of weather. But in the slope, projecting only a few inches, was the great metatarsal of a *Baluchitherium* nearly two feet long and as thick as my arm.

From this point I could see the opposite side of the ravine. Just above the barren red layer a spot of white appeared. That was something smaller and obviously different. It proved to be the broken jaw and teeth of an *Entelodon,* a giant pig that had teeth strangely like those of a carnivore. Building a little *obo* of stones I left it severely alone for Granger's attention.

Two more ravines yielded nothing, but the third produced half a pelvis of our *Baluchitherium*. And what a pelvis it was! Larger than a bass drum! I left that too, for it was badly broken. An hour later a great cervical vertebra beautifully preserved showed on the side of the yellow cliff-face where it hung precariously balanced on a rotting ledge. A few more gales would have cut away its crumbling foundation and sent it crashing to the bottom of the canyon two hundred feet below. With difficulty I hobbled over on my wounded leg and dragged the huge bone back to a place of safety. It probably weighed fifty pounds. With the mass of muscle and tendons and the other six vertebræ, I could imagine what the neck alone without the head would weigh! Thus ended my morning's hunt. Other mornings followed, usually filled with as great success.

The night before two drunken lamas had called at camp to inform us that there were to be great festivities about the *obo* and that we must not work in the vicinity. I said little for the priests were too much under the influence of Chinese wine to comprehend. But Granger and Thomson had a beautiful skull of a large titanothere (*Embolotherium*) partly excavated about a quarter of a mile from the *obo*. We had no intention of abandoning it and at six o'clock that evening they brought it safely into camp. It was an enormous specimen, thirty-seven inches long; with great nasal bones which projected at a sharp angle into the air.

On Sunday, June 24, Young and Horvath started for the broken car guided by a map which Hill had prepared. The rest of us went into the badlands early. The slopes of the top formation are so abrupt that it seems unlikely that a skeleton of *Baluchitherium* would have been preserved. The animals must have been extraordinarily abundant for broken fragments of the huge bones are scattered over the floors of most ravines. Hill, Perez and I discovered a fine cervical vertebra and the distal end of a humerus. I left Hill shortly before noon and a few moments later he discovered the skull of a *Baluchitherium* lying palate uppermost just above the rim of a perpendicular wall. It was in bad condition and there was grave danger that it might slip off into the canyon for the matrix was of worn, yellow sand. Thomson did an exceedingly skillful job in bracing it up. The teeth and front of the skull were gone but the top and occipital parts were of great value. The next day Shackelford made an astonishing discovery.

He came into the mess tent for tiffin and casually remarked that he had found a "bone." Rather too casually, I thought. I was sure that the half had not been told. After suitable encouragement he admitted that it was a large bone—a very large bone. Only the end of it was projecting from a hill slope, but that end was as big as his body. There was a roar from the table at that, for Shack's body is far from thin. He is not exactly globular, but he certainly is fat. A bone as big as any part of his torso would be some bone.

"Don't believe me then," quoth Shack, "but I'll show you."

And show us he did. Walter Granger, Bill Thomson, Shack and I went there in a car, for the place was two miles from camp. It proved to be a gray slope which dropped off abruptly into a deep ravine. Ten feet down the side lay a great white ball. Until I examined it I would not believe that it was bone, for it actually was as thick as Shackelford's body. A little brushing off of yellow sand showed it to be the head of a humerus, or upper arm bone. More brushing exposed its entire length and brought to light the end of another massive shaft which ran deep into the hillside.

All of us stared in amazement. It was not easy to ruffle the calm of Granger and Thomson. They have been at it too many years and have dug up too many strange beasts. But they got a real jolt when they saw those bones. As for me, I was too impressed even to talk.

We supposed that they represented the *Baluchitherium,* the colossal rhinoceros of which we had found a skull in 1922. That beast was bigger than the most gigantic mammoth. It was the largest mammal that ever walked upon the earth so far as science knew. Drs. Gregory and Granger estimate it to have stood seventeen feet nine inches at the shoulders and thirty-four feet in length from nose to tail. Still these figures mean little unless you can visualize them. Just pace off your room and estimate the height of the ceiling; then you will get some conception of the size of a full-grown Baluch.

The upper arm bone which Shackelford had found was as thick as a man's body and four feet long. A man's humerus would look like an unimportant sliver beside it. And remember that it was the shortest bone in the beast's fore limb above the feet. The second giant shaft proved to be the radius. It was five feet long and so heavy that two of us could hardly lift it. In order thoroughly to prospect the deposit, the side of the hill must be removed; it might reveal an entire skeleton.

So at seven o'clock the next morning half the men of the Expedition were shoveling energetically at the coarse yellow sand. The bones were so hard and big that there was little risk of breakage; therefore Granger allowed me to work around them with a curved awl and a whisk broom. Usually I am banished from the immediate vicinity of an important specimen. I can find fossils right enough, but my impetuous nature is not suited to the delicate operation of removing them. I simply cannot work for hours or days, as the others do, before I even know what is there. My pickax methods do get quick results, but they are a bit rough on the specimens, I must admit. In the language of the Expedition, when a fossil is broken beyond repair it has been given the "R.C.A."

Working out a prospect is always fascinating; if it happens to be an unknown beast, it becomes a thrilling adventure. Lady Evelyn Carnarvon once told me of her feelings when she peeped for the first time into a chamber of Tut-ankh-amen's tomb. Hers were the first eyes that had looked into that room since it was walled up four thousand years ago. She could not have been more excited than I was as I brushed away the golden yellow sand that had inclosed our specimen for six million years. The tomb of the Egyptian king gave a glimpse of the world of men and their way of life when civilization had only just been born. Our glimpse was of an incredibly more ancient past, millions of years before man had come upon the earth.

Before the massive radius lay bare for its entire length

Granger discovered another from the opposite side; also two enormous ribs. Just behind them, farther in the hillside, my brush exposed a corner of a flat bone; then a huge tooth nearly as large as an apple came into view. That gave all of us a thrill, for a skull with teeth would have been the last desiderata. But it proved to be only a jaw, and the left side was gone. The doctor next uncovered the middle metatarsal of one foot. A human metatarsal is about four inches long; this one is nearly two feet in length and larger than a rolling pin.

Then we paused to have a look at things. The shoveling squad had removed fifteen feet of hillside, leaving a flat bench where the bones lay exposed. They were all on the same level, close together, and the ends pointed in the same direction. It was obvious that the deposition had taken place in the bed of a swift stream flowing north. Cross-bedding of the yellow gravel and the position of the bones told the story. The animal had died in the stream, the flesh decomposed and the skeleton disarticulated. The smaller parts had been carried on by the water; doubtless many had been broken by pounding against rocks. The massive limb bones had been left where the beast died. They were too heavy even for a torrent to move more than a few feet.

It was useless to dig farther into the hill, for we were rapidly getting out of the stream bed. Only excavations along the watercourse northward would yield results, but unfortunately a deep ravine had cut through it in that direction and the ancient bed was gone. We were disappointed not to have found a skull, but the jaw was some compensation.

The *Baluchitherium* was just about as big as a land animal could grow. Nature has put a very definite check upon size. If an animal grows too large it cannot move about readily enough to obtain sufficient food. Neither can it adapt itself to any radical change of conditions, such as climate, which affects food supply. The inevitable result is the extinction of the species. Baluch browsed on leaves from the branches of trees. When conditions changed and the forests began to disappear he and all his large relatives died, because they could not get enough to eat. He never got to America, for he was much to big to make a long journey. Neither did he reach Europe. Central Asia and Northern India appear to have been his playgrounds.

Shackelford was immensely pleased with himself and his great bones. He had good reason to be too. One does not discover the world's biggest mammal every Tuesday and Thursday! After such an event one has the satisfied feeling that he has not lived in vain. But we did not let him enjoy his peculiar distinction for very long. Heavy competition began at once. Captain Hill found a huge skull that represented the same species. Then Shack matched it with another not quite so good. The doctor hunted early and late and produced an assortment of valuable bones of our titanic beast. I trailed along with three or four vertebræ and parts of feet. Altogether they give us a pretty respectable representation of the skeleton. Of course most of them came from different individuals, but for scientific study and description that is not an insuperable difficulty. A paper restoration has been made at The American Museum of Natural History by Drs. Gregory and Granger.

The marvelous badlands which lay just below our camp produced other new beasts. One of them while not as big as the Mongolian colossus was even more extraordinary in appearance. It belonged to a group of mammals known as titanotheres, which superficially resembled rhinoceroses, but had no direct relation to them. They became extinct many millions of years ago and are not represented among modern mammals. Until we went to Mongolia it was supposed that America had a corner on the titanothere market.

None had been discovered elsewhere, with the possible exception of a doubtful fragment from Austria. But Professor Osborn had been studying titanotheres for twenty years and he was certain that they did not originate in America. He believed that they must have come to us out of Asia.

One of the last things Osborn said to me when I left New York in 1921 was: "Look out for titanotheres. You surely ought to find them in Central Asia." And find them we did, just four days after reaching Mongolia. The first ones discovered were so close to some of the American species that they could hardly be distinguished. It was a remarkable demonstration of Professor Osborn's scientific reasoning.

Since that time we have found a dozen remarkable species of titanotheres. But the most extraordinary and grotesque of all was the one discovered just below Baluch Camp. It was named *Embolothere andrewsi* (Andrews' battering ram beast). The creature was much larger than the largest living rhinoceros. Its skull is concave and shaped like a Western stock saddle, the occipital ridge corresponding to the cantle and the nasal bones to the pommel. The fused nasals project straight up at right angles to the skull and swell into great bulbous ends. He carried his nose in the air if ever an animal did.

Nature seldom does a thing without good reason and probably a vertical nose served some useful purpose to the beast in the beginning of its development. But in this case evolution seems to have run away with itself, as it sometimes does; it got started and could not stop. The enormous antlers of the extinct Irish elk are an example, as, indeed, are those of our American moose. Although at first moderate-sized antlers were useful to a moose, certainly in the living species the development has long since passed utilitarian limits; now they are a distinct disadvantage in size, weight and in the strength it requires to renew them annually.

I suspect that our new titanothere was a browser and fed on leaves from bushes and low trees. In that case he must have had a short thick trunk or pendulous upper lip to bring the front of his face into working position. This beast represents an entirely new phylum or branch of the titanotheres. It is doubtful if they ever reached Europe or America, as nothing even remotely resembling them has been found there. We have seven of the huge saddle-shaped skulls, as well as many other bones of the skeleton.

The chaos of ravines and canyons below the camp seemed to be full of important specimens. It took only an hour or two of prospecting to find something new.

It was easy because our men knew their jobs. Yet fossil hunting seems to be rather a mysterious business to the uninitiated. I am asked at least a thousand times a year, "How do you know where to look?" We are regarded as having some magic power to be able to go into a new country and find bones that have been buried for millions of years. But I can assure you that there is nothing occult about it. Walter Granger is one of the best fossil hunters in the world and he is not a bit occult. I have lived with him in the same tent for many years and I ought to know. He is a tall Vermonter with blue eyes, brown hair, an infectious smile, and is the best-natured man I know. His fossil-hunting equipment consists of a profound knowledge of prehistoric mammals and geology, of keen eyes, unlimited patience and long experience. His success is due to his natural gifts plus careful training. Albert Thomson is another man very much like him. Personally I am a rotten fossil hunter; my temperament is all wrong. But it is great fun to mess about when I have the time and sometimes hit upon a good thing. I have tried almost every kind of sport and nothing gives me a greater thrill than developing a new fossil.

Where to look is simply a matter of scientific knowledge. Fossils can be found only in sedimentary deposits, such as a sandstone, limestone, clay and slate. Of course bones could not be preserved in igneous or volcanic rocks, such as granite. Therefore we travel over the country until we find sedimentary strata. That is the first requisite. Next, the deposit must be cut by erosion into escarpments, ravines, gullies or canyons. Thus a cross section of the strata is exposed, and if bones are lying far below the surface, ends or bits are likely to show. It becomes, then, merely a matter of trained eyes to find them. Sometimes only the tiniest fragment will give the clew to an entire skeleton; but always there must be some such indication of the presence of bone. It would be quite hopeless to dig just anywhere. It is an axiom in palæontology: "Never dig for bones unless you see them."

Not all sedimentary strata contain fossils. Often there are great areas of beautiful badlands which are absolutely barren. In the very top layer of yellow sand at our camp we found many bones of the Mongolian colossus. Below it lay sixty feet of brick-red sediment; lower still was a gray-white stratum, then more red, and at the very bottom another layer of gray claylike sand. In the two red deposits hardly a trace of fossils existed, but the gray layers were rich. It simply means that when the red sediments were being deposited, there was a period of considerable aridity, and conditions were not favorable for the preservation of bones. Sometimes the sediment in which fossils are found have been consolidated into hard rock, such as sandstone or limestone. Consolidation depends upon pressure, moisture and chemical factors. Sometimes even very old sediments have not been consolidated into rock and still remain as rather loose sand which can easily be dug or brushed away. We found one beautiful skeleton of a small dinosaur in a ledge of hematite; the bones themselves had become completely impregnated with iron and resisted our strongest tools. On the contrary, a nest of dinosaur eggs, which are eighty or ninety million years old, could be brushed out of sand that was almost as loose as on the day when it was deposited.

I have described only two discoveries because those two new beasts were so strange and so huge that they gave pause even to our most hardened fossil collectors. But do not believe for a moment that those were the only specimens which this rock deposit contained. Far from it. Every day some one came in with a new discovery which was less spectacular but hardly less important than the big fellows. Several remarkable types of rhinoceros came to light. One possessed a skull that at first sight resembles an enormous weasel. It is quite new to science. Then there are new carnivores, small hoofed animals and rodents. But the life of that far-distant age was dominated by the huge mammals which I have described.

The thing which surprises us most is that we have not discovered the ancestor of all the horses; I mean the five-toed fellow. Four-toed horses have been found in the Eocene of Europe and America, and there is a pretty complete evolutionary series from that dawn period right up to the living horse. But it is certain that a five-toed ancestor existed, and we believe equally certain that Asia was where he was born. We confidently expected to find him in Mongolia. There are half a dozen good reasons why he should be there. Yet we have discovered just about everything else and not one trace of horses in the ancient formations. It is not until near the end of the Age of Mammals, the Pliocene, that horses suddenly appear. Where were they if not in Mongolia? The only answer appears to be that their evolution took place farther to the north and that somewhere in southern Siberia their remains will be found. That ancient continent, known as Angara, will doubtless yield most important results, once it is investigated palæontologically. It is quite possible that it will give us man's ancestors as well as those of the horse. It is logically the next place for us to explore. I would love to do it if we could. But unfortunately politics and palæontology do not seem to get on well together. Hunting fossils, which involves geological studies and messing about in the earth, is too easily confused with oil and mineral research by suspicious and ignorant politicians.

CHAPTER XVI

End of the 1928 Expedition

WE had a good deal to talk about in camp besides fossils and science. The radio furnished some of the conversation. As a rule I am not a believer in radio for an expedition such as ours. I find that the entire personnel is happier, more contented and works better if there is absolutely no news from the outside world.

Most of the men have wives and families. We leave them in good care, confident that all will be well. But suppose that a man gets word, when we are a thousand miles out in the desert, that his wife or children are dangerously ill! It is impossible to return or, if he could, not for weeks or months. Anxiety would ruin his work. Better that he knew nothing, since he could be of no assistance.

But 1928 was somewhat different, since political conditions in China were extremely uncertain; moreover, we needed the correct time for longitude observations. Our set was for receiving only. I arranged with the American Legation in Peking to send us any really important news and with the U. S. Navy to transmit time signals on a short wave from Cavite, Philippine Islands, each night.

The Mongols, of course, thought it a miracle when we let them listen with the head phones to broadcasting from the theaters of Vladivostok and Khabarovsk in Siberia. It seemed like a miracle to me, too, out there in the desert where one has time really to think. We could even hear the sound of feet on the platform and the talking of the audience between acts. It thrilled me always when the time signals came in, for that was something direct and personal. I know the naval station in Cavite. The moon shining across the waters of Manila Bay on palm trees and flowers; the stifling heat; the operator dressed in white at his keyboard. At ten minutes to ten:

"Here goes for those fellows up in the Gobi Desert."

Out into the night he sent the warning dots and dashes followed by the signal at precisely ten o'clock. I wondered if he ever tried to picture us as we received his message: The long semi-circle of blue tents; the restless mass of kneeling camels; the grim reaches of the soundless desert. All under the same moon that looked down upon him in his tropic garden.

At Baluch Camp the Expedition had begun to resemble a traveling zoo, for it swarmed with young birds and animals. Shackelford rejoiced in the possession of two baby horned owls, which we had taken from a nest in a beautiful canyon south of camp. Horvath had a pair of ravens; Doctor Perez a beautiful falcon and a kite; Granger two golden eagles; Mac Young a gazelle, and Buckshot a hedgehog. I was the fond possessor of a young duck caught by Doctor Perez. There was also a Mongol puppy and my police dog Wolf.

All these birds and animals assumed a really important place in the Expedition. Out of touch with the rest of the world for months, as we are, anything that will give a new interest, which will take our minds off our work and each other, is valuable. It assists in keeping up the morale of the whole party. That is the reason why I am willing to sacrifice space and weight to our Victrola. It is a mental relaxation and a thing which helps to keep our sense of proportion. More than once, I have seen a man who was irritable put in good humor again by listening to a record of the "Two Black Crows" or watching the antics of one of our pets. Such things are wonderful stabilizers.

Shortly before we left Baluch Camp, Shackelford made another discovery. He was continually poking about in odd places where no one else had prospected. During one of his rambles a few fragments of bone in the bottom of a shallow ravine caught his attention. Following up the line of the wash he saw other pieces embedded in the earth. Excavations revealed a pelvis with the hind limb in place of what is doubtless our gigantic Baluch. The bones were in the side of a small ridge formed by two diverging gullies. On the opposite slope parts of the animal's fore limbs were exposed. The great beast had died lying on its right side. As the bones are in their proper relative positions it is highly probable that the entire skeleton is there.

Unfortunately the surrounding matrix is of tough rubberlike clay. It can only be picked off bit by bit, and the bone itself is so soft that it must be continually hardened with shellac. It would require weeks of time and great quantities of material to remove this colossal skeleton. We had neither. There was nothing for it but to recover the bones with earth and obliterate all traces. They will remain in safety until we can return properly equipped with special materials.

Although the badlands at Baluch Camp continued to yield new specimens every day, I was anxious to be away.

End of the 1928 Expedition

The other members of the Expedition had finished their work. Only the palæontologists were busy. That year we were equipped to explore new country; the next season would be devoted to intensive work. Our next camp was on a beautiful red mesa. To reach it we crossed a great basin. In the middle was a fossil forest. Tree trunks, branches, stumps and slabs of wood lay scattered as they had fallen there millions of years ago. It was doubtless on the leaves of these same trees that our new titanothere, he of the uplifted nose, had browsed.

The mesa was approachable with cars only at one point. Geologically of the same formation as the one we had left, we supposed it would contain similar fossil mammals. In proof we found an *obo* at the base constructed chiefly from bones of our *Baluchitherium*. A huge *obo* surmounted a detached pinnacle at the southeast end of the mesa and a whole family of baby *obos* reposed below it. Therefore we assumed that there would be objections from the lamas if we remained there long.

The Mongols have no religious superstitions concerning fossil bones (at least we have encountered none) but they do dislike to have the ground disturbed in the region of their *obos*. Apparently they believe that spirits dwell therein. Of course we respect their wishes. Still it is nothing short of miraculous how tractable the spirits become, if a few dollars slip into the lamas' hands. This particular mesa was deserted. Beautiful camel feed flourished and died uneaten on the tablelike top. There was not a sign of human life or of winter camp sites. The priestly visitations began with scouting parties on the lower plain. Then a lone lama visited us and was received with courtesy. Next, ten dirty blear-eyed priests arrived. The mesa, they said, was holy ground, extremely holy; so holy, in fact, that they did not even pasture camels there. I have forgotten just how much was required to pacify the spirits; I think it was a dollar for each lama.

The Holy Mesa produced several vertebræ of our new colossus, larger than any in our collection; also the hind limbs and feet of a smaller individual. The jaw and teeth of a gigantic new rhinoceros and a dozen smaller things were added in quick succession.

But the lamas seemed not to have entirely appeased the spirits of the Holy Mesa. The hottest weather I have ever known in the Gobi held us for two weeks. During the day the thermometer hung at + 110° F in the tent; in the sun it reached + 145° F. The change between night and sun temperature was seldom less than 70° F. A deluge ended the great heat, but it had worked havoc with our gasoline. All the gas had been packed in special tins and cases, but the enormous temperature changes had been too much for them. Tin after tin burst along the solder lines. Nearly a thousand gallons of our precious gasoline disappeared into the air. It meant that our running capacity was reduced just that number of miles because the fleet of eight cars used a gallon to a mile. We have not yet solved the problem of how to transport gasoline on camels. Steel drums would not leak, but there are half a dozen reasons which make their use impracticable for our work.

From the Holy Mesa a drive of a hundred miles brought us to Iren Dabasu, or Erhlien, a salt lake where we had found the first strata of the Age of Reptiles, in

1922. Work there the following year had given us a superb collection of dinosaurs: the duck-billed Iguanodon type of Europe, carnivorous dinosaurs, and a little upland fellow that could run on his hind legs like an ostrich. It was here that we had had the first confirmation of the theory that Asia was the mother of the life of Europe and America. Early in the spring of 1923 we had found some peculiar egg-shells at Iren Dabasu. At that time we did not connect them with dinosaurs. Later the same year, after the now-famous dinosaur eggs had been discovered, we began to think about those shells. Eventually they were submitted to Prof. Victor Van Straelen of Brussels, a noted authority on the micro-structure of eggshells. He pronounced them to be undoubtedly dinosaurian. Moreover, they were of quite a different type from those found at the Flaming Cliffs. Thus we had located the second place in the world yielding dinosaur eggs.

Since 1923 we had been unable to return to Iren Dabasu. In 1928 we went there egg hunting. On a low gray ridge fragments of shell were immediately discovered, and Granger decided to do a bit of excavating. Hardly two feet under the surface he found several nests close together. Evidently the hen dinosaurs regarded this porous sediment as particularly suitable for egg hatching. They were the duck-billed iguanodonts, which sat upon their hind legs and used the short, weak fore limbs only in feeding. Imagine this sand bank fifteen or twenty millions of years ago crowded with dinosaurs. Each one scooped out a shallow hole in which to deposit its eggs. These were nearly round, hard-shelled and probably white, fifteen to twenty-five in number, arranged in a circle with ends pointing inward and in two or three layers. After the mother dinosaur had covered them lightly with sand she left them to be hatched by the sun's warmth. But doubtless each one kept a watchful eye upon her nest.

That sand bank must have been as popular as a bathing beach in summer. Only a short distance away was a lake margined with lush vegetation. The region literally swarmed with dinosaurs. Although we have found the remains of hundreds of individuals, think how comparatively few of the actual number would be preserved as fossils; still more, what an infinitely small proportion of those preserved will ever be discovered.

These new eggs differ considerably in shape from those of the Flaming Cliffs. They are much less elongated and somewhat resemble very large crocodile eggs. The entire hillside seemed to be full of shell, but we discovered only five nests containing even fairly presentable eggs. Certainly further excavations would yield many others.

Our appetite for eggs was satisfied in two days and we proceeded east of Iren Dabasu. We know of only one man, Campbell, who has crossed that region and left any written account. A month earlier, on reconnaissance, when our car had broken, we had had a look at some of it and located several fossil deposits. One of them was a gray escarpment bordering a vast flat basin. We found the intervening country to be typical of the Gobi—hard gravel plains, sage-covered depressions, gently swelling hills. The escarpment itself offered a wonderful view. It was like being on the deck of a ship and looking out upon a tranquil yellow-green sea.

The actual surface of the bluff on which we were camped was the fossil-bearing layer. Moreover, by little else than luck we had stopped at the richest spot in the formation. The first discovery was a mastodon skeleton. The beast seemed to have died all over the place. His skull was found near camp; twenty feet away was the lower jaw; a few yards on either side, fore and hind limb bones; also several ribs in another pile. Evidently some carnivorous animals had dragged the great carcass about before the flesh was gone. We named the place Mastodon Camp.

In the same stratum there were great quantities of clam-shells; layer upon layer of them. It did not take our scientists long to realize that we were on the edge of what had been a vast fresh-water lake. The limits were easily definable; also, that the exposure was much younger than any we had found that year. It belonged to the Pliocene. All of which was exceedingly important. Everything we found there would be new and would throw a brilliant light upon what had been the darkest period of Mongolian life history. Moreover, it was just the geological age in which we might look for early human types with greatest hope of success.

As the mastodon skeleton was freed from rock it developed that it was a very young individual, almost a baby, but a most extraordinary specimen. The skulls of almost all the Proboscidea—mastodons, mammoths and the living elephants—are extremely short and wide. This one had an elongated narrow rostrum. Embedded in it on either side were two slender tusks. Except for the teeth and the tusks one never would have guessed that it was a mastodon.

A few days after the find Captain Hill brought in two flat plates, each about eight inches wide by nine inches long but only half an inch thick. Enamel and dentine showed them unquestionably to be teeth. But what sort of teeth, how they functioned, and what animal wore them was a mystery. We were completely defeated. Later several more were found; also an enormous shoulder blade. Evidently some extraordinary creature, unlike any with which we were acquainted, had lived here during Pliocene times. Every day Granger and I would look at those teeth and try to imagine to what they belonged.

We did not learn until two weeks later. Then, in another deposit of the same formation thirty miles away, a pure accident gave the explanation. The last evening before breaking camp, Granger was walking back to dinner. He climbed up the escarpment to the plain on which the tents were pitched. Two feet below the edge he stepped on a fragment of bone. There was more scattered about and a large piece firmly embedded. After a little excavation he realized that he had solved the mystery of the flat teeth and returned to camp to get me. What he had discovered proved to be the lower jaw of a mastodon similar to the young one found at Mastodon Camp but much larger. The spatulate front of the jaw can only be described as resembling a great coal shovel. Side by side horizontally in the end were two flat teeth like the others we had found. They were eighteen inches across. Behind them the jaw narrows and then divides into the two branches which bear the molar teeth. The jaw is more than five feet long—a perfectly stupendous organ.

It exactly resembles a scoop shovel and I believe that a scoop shovel it must have been. The fact that we found many mastodon remains on the very shores of the great inland lake gave a clew. Quite probably there was much lush vegetation there. Our mastodon must have waded along the edges scooping up the water plants with his great shovel. By means of a trunk or tongue he pushed them into the back part of his mouth to be masticated by his molar teeth. Without doubt the huge shoulder blade that we found at Mastodon Camp belonged to one of the full-grown mastodons. Our first skeleton was that of a very young individual. The flat teeth had fallen out of its lower jaw, for they were rather loosely set in; so we did not recognize the affinity to the plates which we had been finding.

At Mastodon Camp one of the Chinese collectors discovered the almost complete skeleton of a rhinoceros. It was about the size of one of our living rhinos and lay in what had been a stream bed. When the bones were exposed it resembled the skeleton of a horse or a cow as one sees them so frequently on the desert; so much so that I could hardly realize that it had been buried for millions of years.

During the stay at Mastodon Camp, Captain Hill, Mac Young and I had gone to a mission station one hundred and forty miles from Kalgan to leave six fossil cases too large to be carried by camels. Our route was through new country, which we mapped and explored. At the mission we heard of Chang Tso-lin's death by a bomb, and other important news. The entire political situation had changed about as completely as possible since we left in April.

During our absence from Mastodon Camp the camels arrived. The gasoline situation had become so serious through leakage that it was evident our work would have to end soon. I determined to concentrate on exploring the new country about us and carry all the remaining gas in the cars. The camels were started for Hatt-in-Sumu.

The next three weeks were of great importance geographically. Still they did not present features of unusual popular interest. A general survey of the country and mapping were our chief occupations. We located a great area of dead sand dunes which extends north and south for many hundreds of miles; found a dozen small lakes and new geological strata; also a great residence site of the Dune Dwellers.

By the time our gasoline was at an end we had completed our exploration program. Our fossil collection numbered ninety cases; we had ten thousand archæological specimens; the geologist had discovered half a dozen new formations and had studied a vast untouched area; the topographer had mapped three thousand miles of blank space on Mongolia's map. We were well content.

The American Legation, at my request, had notified the officials along our return route. Thus we encountered none of the expected difficulties. But our entry into Kalgan very nearly resulted in a tragedy. We arrived in black night and heavy rain. Four miles outside the city the steep clay hills were like grease. Our cars could make no progress. While roping a wheel, Horvath, one of our motor experts, drove a knife into his thigh, severing an artery. The blood pumped out in great jets. He would have died in a few minutes had the surgeon's car not been near his own. To manufacture a tourniquet in the mud and darkness was difficult, but Doctor Perez finally stopped the bleeding. Horvath was soaked with blood and very weak when we reached Kalgan but a few days put him right again.

When Chang Tso-lin left North China he took with him all the cars and locomotives that he could find. As a result, Kalgan was connected with Peking by a train that ran sometimes once a day, sometimes once a week. It was useless to think of getting our motors and equipment to Peking by rail. Either they must remain in Kalgan indefinitely or we must drive them down. We chose the latter course. Driving one hundred and twenty-four miles does not seem on the face of it to be a very serious undertaking. But to drive that distance over a road that was virtually a swamp and through a mountain pass where the trail was only designed for mules and camels is what we had to do. Moreover, our cars were loaded heavily.

It was three days before we reached Peking—three days of fighting the road from daylight until after dark. At six o'clock of the first evening the trail had led us into a wide dry river bed. Suddenly we heard a roar from the hills to the right and saw a brown flood sweeping out from a narrow valley. In a few minutes a wild torrent waist-deep was across our path. To get out while the getting was good was obviously the thing to do. Just before we reached the bank from which we had come there was the same ominous roar, and again we were cut off from safety by a wall of water. It looked distinctly bad. We drove to the highest part of the river bed and watched the water slowly rise toward us. Finally I could stand it no longer and we made a desperate attempt to reach the shore. The first car dropped into a hole, breaking an axle. The next car got through. The third hopelessly sank in the mud. We worked feverishly all night and before daylight got all but two of the cars across. Then the water began to recede rapidly. Had we remained where we were we should have been all right; but one never knows.

The Nankou Pass, where the Great Wall was built to keep the Mongol raiders out of Peking, very nearly kept us out also. The road was such a nightmare of rocks and boulders that I shudder even now to write about it. At any moment I expected to see some of the cars crash. How anything on wheels designed to travel on a road could stand the punishment those Dodge cars received was beyond my conception. But they kept steadily on and at long last we came out on the road at the foot of the pass. Peking was only thirty miles away. We passed through the great gate of the Tartar wall at eight o'clock. The Expedition of 1928 was ended.

Almost immediately trouble began with an unofficial body of Chinese called the "Cultural Society." Later this became what is now the "Commission for the Preservation of Ancient Objects."

Our specimens were confiscated by the Cultural Society on the grounds that the Expedition had "trespassed upon China's sovereign rights"; that we had "stolen

China's priceless treasures"; that we "were spies against the Chinese Government"; that we had been "searching for oil and minerals."

This, in spite of the fact that since 1921 the Expedition had been carrying on its work in coöperation with several Chinese scientific bodies; and that the results of its work had been spread on the pages of almost every newspaper of the world!

Eventually our specimens were released but bitter feeling was engendered on both sides. Negotiations extended to Colonel Stimson, Secretary of State of the U. S. A., the Chinese Minister at Washington and the Minister of Foreign Affairs of the New Nationalist Government at Nanking. I shall not give details of the wrangle for it is best forgotten. In the last analysis anti-foreignism underlay all the trouble; that, accompanied by the increasing nationalistic spirit throughout China. Any agitation of whatever character that was directed against foreigners found immediate popularity with the masses.

It is probable that if the same situation arose today the Chinese would realize that the wisest course is to conform to long-standing international precedents in scientific work.

The entire year of 1929 was devoted to costly and nerve-wracking diplomatic negotiations and it was not until 1930 that the Expedition was again able to take the field.

CHAPTER XVII

The Mysterious Lake

REPORTS of a mysterious lake continually came to us when the Central Asiatic Expedition was in Mongolia during 1928. Natives said that its shores were covered with fossil bones; that it was far out in the dead sand dunes to the east, in a region so desolate that few Mongols had ever been there. The name was Tukhum Nor.

Native information is proverbially inaccurate, but we heard the tale so often that we decided to investigate. We sent out a Mongol who knew something about fossil bones, with instructions to find the lake and bring us specimens. He went on a pony, but lost himself in the trackless dunes. After wandering with little water and less food, he reached a Mongol *yurt*. He had had more than enough for that year. The next summer we started him off again on a camel, and that time he found the lake.

"There are many dragon bones," he reported. "Oh, a great many. The shores are covered with them. But it is a bad country. All yellow sand and no wells." To confirm his statements, he produced a bag of specimens. We identified giraffe, horse, rhinoceros and other species indicating a Pliocene formation—that is the geological period just preceding the Pleistocene, or Ice Age. Such strata were more interesting to us than any other, because in them we might hope to find the bones of primitive humans. "Hope," I say, for we never "expect" to find primate remains. It is too big a gamble for confidence. All we can do is to work and hope.

I organized the 1930 Expedition with the region of this lake in view. In the heavy sand we would have to travel by camel, and we took only four cars. We were all set for a howling desert—dozens of water bags, huge wooden casks and all our gear arranged for caravan transport.

This I did during the winter in Peking. That was the easy part of it. The diplomatic negotiations were an entirely different matter, as I have already related. Then, when the diplomatic details were arranged, several Chinese generals in the north revolted against several Chinese generals in the south. That threw another wrench into the Expedition's machinery. Having spent the entire winter trying to satisfy one party, it had to be done all over again with the new régime which controlled the north.

Even when I cabled Granger, Young and Thomson to leave New York for China by the first ship, I was by no means certain that we could go. It was then early April and the Expedition ought already to have started for the Gobi. By the time the men reached Peking, in May, I had promises of all the permits. Having obtained them from both sides, we felt reasonably safe.

But there was still another little matter to be adjusted. Getting the equipment and cars away from Peking and into and out of Kalgan, our starting point, put a permanent crimp in the Expedition's finances. Eight tax offices in Kalgan alone took their toll. By the time they had finished with us I felt like a margin speculator after the stock market crash. Still, some one has to pay for the fun the Chinese generals have in making war upon one another. And I don't mind stating that the Central Expedition did its share. They ought to be able to conduct several wars on what we had to contribute to the tax officials that year.

We left Peking on May twenty-sixth. The camels, food and equipment awaited us at Hatt-in-Sumu, on the great plateau, one hundred and forty miles north of Kalgan. The camels were a majestic-looking lot. That is about the only word of praise as to his personal appearance that can be applied to a camel. He is majestic, but handsome he is not. He was made up of spare parts when the Creator had used everything else for regular animals.

Camels don't like white men. Apparently our body odor is extremely distasteful to them, and they evince the fact by spitting upon every visitor. To be spat upon by a camel is a joke only to the onlookers. The spittee feels very differently about it. Fortunately you can see it coming in time to run if you know the symptoms. When a suspicious distention appears at the base of the long throat and begins to travel upward, a camel-wise man leaves the vicinity at once. We veterans of the Expedition take a malicious delight in having new members initiated. Doctor Garber, our surgeon, got it first that year. In the performance of his duties he had to inspect the noses of several camels. One of them gave him such a bath of partly digested vegetation that he was washing himself for the best part of a day and even then couldn't get rid of the smell.

The Mysterious Lake

Our Mongol guide awaited us at Hatt-in-Sumu. He had reached the fossil-bearing lake from the south through the heavy sand, but had returned by a better route to the north. He believed that there the sand was hard enough to bear a motor. We might get within twenty-five or thirty miles of the lake, he thought. I decided to make a reconnaissance with two light cars before we went in with the whole Expedition. Five of us started out the next day. At first we traveled over beautiful rolling grasslands dotted with herds of gazelles and flocks of demoiselle cranes doing their mating dance. Some Mongols and flocks of sheep, but surprisingly few. Country just like our West of a hundred years ago, minus only the buffalo. No fences, of course. I never have seen a fence of any kind in all Mongolia.

Car trouble stopped us for an hour, and we saw a wolf kill a sheep. The wolf dashed out from the ravine, separated two sheep and tore open the throat of one while the other stupidly stared at its dying companion. The herder ran into view and the wolf snatched a few hasty mouthfuls before it retired. The carcass was left on the hill slope. Mongols are superstitious about sheep killed by wolves. Only the poorest natives will eat one. It is bad luck.

After a few miles, the grasslands ended and sand began. The Mongol was right when he said that it was a desolate country. Utterly God-forsaken, we called it. Just sand—sand basins, yellow dunes, and plains. It was very hard going. Most of the way we had to push the cars on strips of canvas stretched in front of each wheel. The monotony of it was as bad as the work. Just push to the end of the canvas, spread it out and push again, hour after hour. The first twenty miles were the worst. After that we seemed not to mind so much. Perhaps we were too exhausted.

Eventually we reached the lake. During all the hard work we had comforted ourselves with visions of blue water margined by green grass; of breeding water fowl and whispering poplar trees. Reality: A shallow depression of sun-baked mud stained with alkali, surrounded by a waste of dunes and niggerheads bathed in streaming heat waves. To complete the picture, a lone sheldrake sitting disconsolately in the exact center of the basin.

Our Mongol could hardly believe his eyes. He had seen it just after unusually hard rains in the summer, when the depression was full to overflowing. But that is like most desert lakes. Here today and gone tomorrow. One of the western Gobi dried up while our topographer was mapping it!

The Mongol pointed to the south side of the basin, where he had found his fossils. An area of irregular gray hillocks marked the deposit. We plodded across the lake bed, walking on hot, sun-cracked mud and resembling blocks of broken ice in an Arctic sea. Our guide triumphantly indicated bits of scattered bone. He seemed awfully pleased with himself about something. I asked him why the smiles.

"Here are the *lung-gu*—dragon bones," he gurgled. "Many bones, many bones."

My sainted aunt! So this was what we had pushed our way through miles of sand to see! A dry lake of stinking alkali mud, and fossils enough to fill a gasoline box—

no more! The few bones there were had been crushed and broken into fragments before they were deposited some millions of years ago. The stuff was valueless except to determine the age of the formation.

As we inspected the exposure, all of us wondered where the Mongol could have found the teeth and bones which had enticed us there. He explained that it was only after long search that he made his collection; moreover, that he had taken all the good things. We admitted that; there certainly wasn't much left.

After a little, we began to see the humor of it all. It was just another experience with native information. We had had plenty before, and might have known better. Still, the things he had brought back were distinctly good, and there was comfort in having that alibi. It was disappointing—extremely so—but that was about all. Had the entire Expedition wasted several weeks by going there on camels, we could not have laughed so heartily. As it was, we would push ourselves out of the sand and begin again.

Returning to Hatt-in-Sumu, I started the camels northward into a region we had roughly explored in 1928. Soon after, we followed. Five of the men had to go with the caravan. Fourteen rode on the cars. The motors looked like giant spiders, for the legs and arms of our Chinese assistants projected at all angles. They had to glue themselves to the loads wherever they could find a sticking place. We had expected to do but little traveling by car, and the failure of our lake project necessitated a reorganization of the entire Expedition.

The course was set northeast over ground made familiar by our 1928 explorations. The route had been mapped at that time, so that now we could travel rapidly, unretarded by topographic work. The first evening we slept under the stars one hundred and thirty miles from Hatt-in-Sumu. Next day we were at Mastodon Camp. The tents were pitched on the edge of the gray bluff where we had camped in 1928.

We felt that if we followed the escarpment both north and south, there must be other fossil-bearing exposures. We wanted more of the great shovel-tusked mastodon. The jaw was fine, but it did not go far enough. Of course, our great hope was that we might find a quicksand or bog deposit. Such spots acted in the past just as they do now. The most famous examples are the asphalt pits of La Brea in California.

There animals came to drink at the shining pools of water which gathered on the surface of the soft tar. The instant that their feet touched the tar they were held as securely as a fly on sticky paper. Once a herbivorous animal was caught, the trap was doubly baited. Prowling carnivores, particularly the saber-toothed tiger, came to the edge of the pits. Perhaps a fat bison lay there, struggling in the tar. With a snarl, the tiger leaped and himself was caught. I do not know the exact number of saber-toothed tiger remains that have been taken out of La Brea, but it runs into several hundreds. Until a few years ago, when fossil hunters began excavations there, the asphalt was acting just as it had done since the beginning of the Ice Age. At the time Walter Granger last visited the pits he saw a rabbit and a heron struggling in the tar.

Bogs and quicksands are similar traps, although not quite so fatal. Unquestionably such spots must have existed along the shores of this ancient Mongolian lake. They would be our best hope, too, for the remains of primates. Even as early as the Pliocene, several million years ago, primitive humans were more intelligent than the animals about them. They probably knew enough to avoid such places. Still, accidents happen. Men get caught in bogs and quicksands even today. I have had one or two narrow escapes myself.

In a country like Mongolia, search for human remains is pretty difficult. It is just about like looking for the proverbial needle in a haystack. Almost all important human finds have been made in dwelling caves. There were very few caves in Mongolia during Pliocene times, and there are few now. I might almost say that they are non-existent. The early men of this region had to accustom themselves to a life in the open, seeking shelter on the lee side of a bank or under rocks. When they died their bones were washed away or scattered by predatory animals. Therefore our best chance is to find the remains of some unfortunate individual who had sunk in quicksand, or some deposit where a gently flowing stream had carried bones along, finally to bury them in soft sediments. We have found many such places in the past, but the strata in which they occurred were too old.

The day after we arrived at Mastodon Camp, Granger, Père Teilhard de Chardin, the famous French savant, and myself went southward along the edge of the escarpment to see what we could see. We followed the shell lines of the old lake shore, investigating every point where the desert vegetation had been eroded and the gray sediments appeared. Everywhere there were bone fragments. On the side of an outlying hill I saw a bit of white bone three inches long. I touched it gently with my foot, but it did not move. Dropping to my knees, I brushed away the surface soil with a whisk broom. A flat bone was exposed; then teeth appeared. I knew that I had gone far enough, and called Granger.

Under Walter's expert manipulation, the jaw of a mastodon was soon exposed, with the molar teeth and one tusk in position. It was not the scoop-shovel fellow, but quite a different type—a round-tusked mastodon. This was a new animal for the formation. Leaving it for future removal, we drove to a long exposure of deeply eroded badlands which were visible several miles away. There was much broken fossil bone lying on the surface. In one spot we discovered what appeared to be four or five bone hillocks. Vertebræ, ribs and limb bones were just visible embedded in the gray earth. We could see that it followed the shell line of the lake shore. Almost certainly it marked a former bog.

Camp was shifted next day. It was a drive of only eleven miles over a hard gravel peneplain. Just as we stopped the cars on the escarpment a big gray wolf dashed over the edge and across the plain. He had been asleep in one of the ravines below us and certainly made a mistake in coming up to see what the noise was all about. I reached my rifle first and rolled him over in full run at three hundred yards. The tents were pitched on the very brink of the bluff. A weird panorama of red and gray ravines and buttes spread out below us. Far away across the basin the Arshanta escarpment showed as a sheer wall of purple and gold. The plain which stretched away behind camp was almost terrifying in its vastness. We knew that for many miles there was neither water nor Mongols.

One morning the cooks shouted that a pair of wolves were only a short distance from camp. I jumped into a car with two of our Chinese. The wolves were legging it straight for the center of the plain, where the going was a bit rough. The car could only do twenty-five miles an hour and the wolves a little better than that. They pulled away from us slowly until we found a stretch of hard gravel. At forty miles an hour we overtook them, and I got in one shot at nearly four hundred yards. The rearmost wolf dropped flat. Rushing past him, we followed the other, which had a long start. He twisted and turned, always seeking the roughest ground. After five miles I got a bullet in his quarters, but he kept going. For twenty miles we followed that wretched wolf. He would disappear over a rise and lie down in a ravine or behind a stone. After we had passed, he would try to sneak out unseen.

Finally we lost him when the car was stopped by a sand pit. Then I looked about to see where we were. For the life of me, I couldn't tell. The sun was obscured by heavy clouds. The horizon was a flat line, except where patches of mirage piled up visionary mountains. There were no more landmarks than at sea. Mentally I tried to remember our course. I had a feeling that we had traveled generally southeast, but in the excitement of the chase I could not be sure. The gasoline tank registered five gallons. Half a bag of water hung on the car, but the radiator was nearly dry. My compass wasn't where it should have been, in my pocket.

Following our tracks backward netted nothing, for we soon lost them on a wide stretch of gravel. We could only travel blindly in the direction where I thought camp ought to be. The horizon line continued flat, no matter how far we went. Now and then I steered toward what I thought was a promontory, only to see it stream off into the vague lines of mirage.

The situation was uncomfortable, but not really serious. If the gas gave out before we reached camp, we would just have to sit and wait. Fortunately I shot an antelope. That would give us something to eat, and we could drink the water from the radiator. I knew that eventually the others would find us. It hardly seemed possible that they would be worried enough to search before evening. In camp they would think that we had joined the party at a fossil pit, six miles south. I would get the car to the highest ground and, when night came, turn the spotlight on the sky. Such a ray can be seen for a long distance.

These thoughts were going through my mind as we bumped our way across the plain. Strange how much rougher it seemed than when we had the wolf in sight. Perhaps I was all wrong and we were going in the opposite direction. I could only trust my instinct, for there was absolutely nothing in the landscape to give a clew. We traveled for two hours. There were two gallons of gas left. That would only give us twenty miles in the rough going. It didn't look so good.

A little later my eye caught a small patch of gray

sediment lying on the prevailing red earth. I stopped and examined it carefully. There was no doubt that it was the fossil-bearing clay of our deposits. There were none of the gray sediments on the eastern side of the plateau; therefore we must be going in the right direction—west. A gallon of gasoline still remained, but my worries had gone. Although the plain appears to be absolutely flat, it really isn't. It rolls away in great undulations like the long, smooth swells of a calm sea. From the crest of a land wave I saw eleven blue masses swimming in the heat waves of the desert. They rose and sank, floated uncertainly in the light breeze, and finally settled to earth as we approached. It was our camp, three miles away. There was still a little gas left when we drove in, but only a little. Another four miles would have consumed the last drop. That indefinable direction instinct which all explorers develop unconsciously had taken us home as surely as a compass. We often marvel at the direction instinct of animals. To me, it does not seem at all strange. That sense is much more highly developed than in man and they have to use it constantly. Mongols have it to an amazing degree—much more than any white man. They use it every day. Without it, they never could exist in the Gobi.

CHAPTER XVIII

The Fate of the Rash Platybelodon

OUR caravan Mongols had taken the camels down into the basin, where the feed was better than on the plain. One of them came up, badly frightened, for a wolf had bitten a girl in the thigh when she tried to drive it away from a freshly killed sheep. He was worried about the camels. Mac Young went down next day and found thirty-three sheep lying on the plains with their throats mangled. Three wolves had rushed into the flock and almost annihilated it in a few moments. The wolves had become a veritable plague. We killed eleven, but there seemed to be many left. In the western Gobi, wolves are remarkably scarce, considering the abundance of game. There we would seldom get more than five or six in a season. We named the place Wolf Camp.

It was a very lovely spot and all of us enjoyed it. Fortunately there were many gazelles on the plain. Antelopes, to us, are what seals and polar bears are to the Arctic explorer. Venison forms the basis of almost every meal and no one seems to tire of it. I do not know any other wild game that we could eat so continuously. I suppose the reason is because gazelle meat has very little wild flavor as a rule.

Every other day I used to ask Lieut. Bill Wyman, topographer, if he wanted to go to market. He never refused and we would start out in the small car. Almost any direction would do. We were sure to find antelope within a few miles. From the summit of a ground swell we would sweep the plain with field glasses until antelope were sighted.

Running toward them diagonally, we could entice them up as a magnet draws steel. A wild ride at full speed across country, rough or smooth alike; a sudden stop and a leap to the ground on either side; two or three shots from each man; usually two antelopes. That is how we would go to market. We seldom shot under three hundred yards, and the animals were running at fifty to sixty miles an hour. It really is not so difficult as it sounds. Often sportsmen have said to me, "You could hardly see an antelope at four hundred yards, to say nothing of hitting it."

But they don't know Mongolian conditions. Crystal-clear air, a flat plain and gazelles running smoothly in a straight line. If they only knew enough to dodge about, we would not get half as many. The first shot usually gives the range, for the bullet always kicks up a spurt of dust. After that, we ought to score with the second and third, and sometimes the fourth; then they are beyond range. Often we have loaded the car with three or four gazelles after only one race. It isn't sport, that is true. But we were after meat. Twenty-five men in camp eat an astonishing amount. Personally I would much rather go to a market and buy my meat across the counter. I've killed too many gazelles—perhaps a thousand—and it has ceased to be fun. I know exactly how well I can shoot, and the only satisfaction I get is in making a clean kill, so that the beautiful animal does not suffer. If it were possible I would not shoot a single one, but it can't be helped. New men always get a tremendous thrill out of hunting antelopes, and it surely is exciting at first. I've had too much of it, that's one reason. The other is that now I am not keen to kill anything except dangerous game. Hunt a tiger on foot and he has a chance to strike back on more or less equal terms.

Wolf Camp proved to be a veritable fossil mine. Our prediction about the bogs and quicksands along the shores of the ancient lake were correct. When the low ridge with fossil exposures just below the tents was opened we found a continuous stratum of bone for thirty feet. The men removed the surface mostly by brushing the sediments away with whisk brooms. The skull of a baby shovel-tusked mastodon appeared. There was appropriate rejoicing in the camp. The next day another skull, and then a series of jaws. Some of them were less than two feet long, but had the milk teeth and flat tusks in position. One skull was that of a just-born infant. The place was a baby death trap.

The matrix indicated quicksand and we could read the story plainly enough. The great mastodon mothers had come there with their babies to drink. Both surely were caught in the treacherous sands, but the mothers were strong enough to extricate themselves. The children, all very young, could not get out. That must have been the case, for with one exception we found no bones of adult mastodons. The skulls of the little fellows showed that all were much too young to be wandering about by themselves. Moreover, there were a good many skulls and jaws of other small mammals—antelopes, deer, fox, and several types of carnivores.

One day, Granger, Thomson and I went to look for a broken skull which Walter had seen on our initial visit to this escarpment. Granger always says that the best way to find a fossil is to lose one. While you are searching for it you are certain to discover something else. The paradox proved true in this case. We simply could not locate that broken skull, but a bit of bone caught my eyes, and, brushing away the sediment, I exposed a mastodon's molar tooth. A few moments later Granger removed a flat stone block. It was like lifting a trapdoor, for under it lay another great tooth, firmly set in bone. Granger followed it down while I looked on, seething with excitement. The enormous spoon-shaped lower jaw of a shovel-tusked mastodon—*Platybelodon*—slowly took shape under his brush. It was five feet three inches long, and the two flat, platelike lower tusks measured fifteen inches across. In the meantime Bill Thomson had determined that the other teeth belonged to an adult female skull. We were exultant, for there was every reason to believe that the specimens were perfect.

What we had found proved to be skulls, jaws and parts of the skeletons of a big mother shovel-tusker and her baby. No other bones were there—only those two individuals. We are at a loss to know just why they died at that particular spot. It could hardly have been a bog or quicksand, for in that event bones of other animals would certainly have been preserved. Perhaps they both fell into a hole or part of a bank caved in upon them. Whatever the cause, there they had died, mother and baby together. It was a kind fate which directed our steps to their unmarked grave.

During the days that we were excavating these two specimens, Père Teilhard de Chardin prospected the escarpment farther to the south. Six miles from camp he discovered an amphitheater in the badlands, capped to the west with snow-white marl. The slopes of the familiar gray sediments were strewn thickly with broken mastodon bones—teeth, fragments of skulls, limbs, vertebræ and dozens of ribs. In a rectangle formed by four blocks of sandstone, part of a jaw was embedded.

The fragments were confined to an area about fifty by thirty feet. There was every indication that it had been another death trap. We began work with the keenest anticipation. Still, no one suspected that we were about to excavate what will go into history as one of the world's most remarkable fossil deposits. The first day's work demonstrated beyond question that this was the site of a former bog. I wish that I were able to give an adequate impression of the thrilling interest in opening this ancient grave. Out there in the desert, in the brilliant sunshine of the year 1930, we were reading the story of a tragedy enacted millions of years ago. Every hour, at first every few minutes, a new page was turned in this book of stone.

We know just what happened. The tale we read is as follows:

A quiet estuary ran inland from the main lake. Lush vegetation lined the shores. Floating plants and green tubers sent their roots downward through the shallow waters into a deep well of mud. A huge mastodon, his monstrous shovel jaw dredging up masses of trailing grasses, worked his way slowly along the shore. The succulent tubers, resting so innocently just beyond the water's edge, enticed him farther and farther into the treacherous mud. Suddenly, amidst his greedy feeding, he found that he could not lift his ponderous legs. He struggled madly, only to sink deeper and deeper into the mire of death. Frenzied trumpeting echoed from the high shores. At last they ended in exhausted gurgles as the colossal beast sank below the surface.

Another came, and still others; each one to die as he had died. Twenty, we know, were trapped; probably more than that. Down in the black mud tons of flesh macerated and dropped away from the great skeletons, leaving the bones to separate, one by one. Some were crushed and broken when other victims piled upon them. Some remained as perfect as when they bore the living flesh.

Perhaps at last the death trap was full to overflowing; perhaps the water, fouled by decaying flesh, sickened the vegetable life and left the trap unbaited; or perhaps the estuary itself dried up. Whatever the immediate cause, we know that as centuries passed into thousands of centuries the great lake disappeared. Countless tons of sediments were deposited on its dry floor. The mastodons' unmarked grave was buried deeply, hopelessly lost.

Then came a change of climate. Gradually the dry winds of the Ice Age, bitterly cold, removed the sediments in the old lake bed. Particle by particle they were swept high in the air, to be dropped upon the plains of China, three hundred miles away. It goes on today. The desert is still giving up its surface, being worn down by the unceasing Gobi winds. Thus it has been eroded for perhaps a million years.

In our search for the hidden stories of Nature, we found the mastodons' grave in that long-dry bog. But we came half a century too late. Already the upper part of the deposit had been worn away by wind and water, which destroyed exposed bones as they destroyed the rocks. Still, much remained. When we removed the cover of sand a mass of fossil bones was disclosed in a thick lens of green clay. They lay like a huge pile of jackstraws. Great scoop jaws, many of them nearly perfect, were heaped upon one another in every possible position. Some extended straight down; others at oblique angles; still others almost horizontal. Most of the jaws were more than five feet long. The part of the jaw which we discovered in 1928 had been enshrined with reverence as one of the prize exhibits of the Museum. There we had half a dozen complete specimens visible at one time. Mixed with them in a seemingly hopeless jumble were enormous flat shoulder blades, pelvic bones, limbs and scores of ribs.

It was difficult to remove any bone, for usually it lay under several others. Only by finding the topmost ones could work begin. The bones themselves were badly preserved. They were like chalk and impregnated with mineral matter only to a comparatively small degree. Walter Granger and Bill Thomson, with their five trained Chinese assistants, did all the work. The rest of us were only too eager to help, but we didn't fit in. After I had dug into a jaw, Granger requested me to cease. I retired to the outskirts, where I could watch operations and drift in fancy back to the days when it all had happened. Doctor Garber fared better, for with his surgeon's temperament and technic he became not only an accepted but a sought-after assistant.

The bones were so soft that as soon as a portion had been exposed it must be soaked in shellac. Thus hardened, more matrix could be removed and more bone exposed. Next, the entire surface was covered with a Japanese rice paper and gum arabic. The paper and gum cemented in place any loose particles and strengthened the specimen. When the entire bone was laid bare, it was covered with strips of burlap soaked in flour paste. In a few hours the paste had dried and the bone was inclosed in a hard shell. Then it could be turned over, pasted on the other side and packed without fear of damage. In the laboratory of the Museum the bandages can be softened with water and easily removed.

It must not be imagained that the deposit was excavated in a few days. Would that it had been possible! For six weeks the men worked there.

The hole assumed gigantic proportions. It was amazing that we did not find more skulls intact. Only three were removed, and they were those of big bull mastodons. But twenty-five ivory tusks were taken out. We wondered why such slender bones as the five-foot jaws had remained unbroken when the skulls were destroyed. Still, an elephant's skull is by no means the solid mass that it appears to be. The walls are filled with air chambers and the skull is like a gigantic honeycomb. Probably they could not withstand the pressure and suction of the mud. Fifteen great scoop jaws were recovered, and half a dozen broken pairs. Virtually all the other bones of the skeleton, as well. There is very little that we do not know about this amazing creature.

CHAPTER XIX

Wolf Camp

WHETHER or not a fossil hunter subscribes to the late Eighteenth Amendment is neither here nor is it there. The fact remains that he must have alcohol. Not for the good of his body but for the good of the fossils. Soft bones must be hardened with shellac before they can be removed. The shellac must be cut with alcohol. The bones of the mastodons at Wolf Camp were like chalk and they just drank shellac. After all, they had been dry for some millions of years.

We had plenty of alcohol, or rather we did have until one night during the end of June, when the quarry had been almost excavated. Then a storm came along that was a storm. About as near a typhoon as one ever gets in the Gobi Desert. Floods of rain and an eighty-mile wind. It was a miserable night. Tents went down on all sides. One moment we lay snug and warm in our fur bags; the next we were enveloped in a sodden mass of billowing cloth, fighting madly to get out. Paddling about, half naked in a freezing rain, in the middle of the night, is either tragic or funny. We chose to believe it was funny. Because every one else was in the same mess, it was not so difficult to see the humor.

Granger and I shared the big tent. The edges had been weighted with whatever was handy—stones, boxes, saddles and bags—anything to keep it down. Next morning, to our intense chagrin, we found that one of the cases we had seized in the darkness had contained our precious alcohol. It was gone, every drop. Leaked out through a hole in the top of the tin.

That situation did not strike us as being at all funny. Work must stop unless we could get more alcohol. It meant going to Peking. Fortunately, that was only four hundred miles. Had the loss occurred during any of the other Expeditions, when we were a thousand miles out in the desert, it would have ruined us.

Mac Young and I were obviously indicated to do the trip. We were not a bit keen about it, for we had not been away long enough to make the fleshpots of civilization desirable. Moreover, the work in camp was fascinating, and Peking in July is like a blast furnace. Since it had to be, I decided to take two cars and load them with specimens. A full-sized war was going on not far from Peking and Mongolia might be cut off by retreating soldiers at any time. The more fossils that were safe in my house the happier we would feel.

Driving away from camp just as the sun rose, we cut straight across the desert to the main Kalgan-Urga trail. The vast, flat plains were glorious in the cool morning air. Hardly ever were we out of sight of gazelles feeding singly or in herds. Bustards and demoiselle cranes circled like aëroplanes beyond the cars. Always there was something interesting to one who could read the book of Nature.

Once on the main trail, we kept a keen watch for bandits. They might take pot shots at us from behind the hills. But the road was free that trip, and we drove steadily mile after mile for fourteen hours. Two days later we were in Peking.

It was the Fourth of July, and we duly presented ourselves at the Peking Club reception. Gossip buzzed. Why had we returned so suddenly? When I told them we came for alcohol, a roar went up: "We thought you were nondrinkers in the Gobi! That's too thin! What is the real reason, now? Just between friends; you know that we'll keep it quiet."

No one would believe the truth. Peking is like that. It scents a mystery, a romance or a scandal in the simplest things. So we let it go. As a matter of fact, most of them decided that I had returned to play in the polo tournament that began that week.

We gazed with satisfaction at the eighteen cases of fossils that reposed safely in our big laboratory. They had been piloted from Kalgan to Peking on the train by Young after passing out quantities of dollars in "squeeze." There is almost nothing that money won't do in China.

A week later we were on the back trail, driving hard for camp. At eleven o'clock in the morning we reached the Black Water, a treacherous swamp, the worst place on the road. Three times in a quarter of a mile our car broke through the thin dry crust and went down to the axles. Only by sinking a foundation in the seemingly bottomless mud and lifting it out with the jack, could the car be extricated. For seven mortal hours we worked feverishly. At this place there are a few scattered Chinese villages of dubious reputation. More than one car had been robbed when mired and helpless. At least fifty people were clustered about, and we felt sure that if darkness found us there we would have a lively night. In daylight we were safe enough, for both of us wore heavy revolvers and kept rifles at hand, even when working at the jack.

By six o'clock we had passed the swamp and were again on the hard trail. It began to rain, first in showers, then in a steady downpour. Darkness came at eight, when we were opposite Chap Ser, the most dangerous place in all Inner Mongolia. It is just a collection of Chinese mud huts and Mongol *yurts* at the extreme outer edge of cultivation. Bad characters, both Chinese and Mongol, gather there, and for years it has been a veritable bandit stronghold. Even though we faced a cold, wet night, there was no question of seeking shelter at that place. We drove past at full speed while hard-faced natives stood in groups, glowering.

Twenty-two miles beyond Chap Ser were the *yurts* belonging to our caravan Mongols. If we could find them in the darkness there would be a dry and safe shelter. Driving over a bad trail in the inky blackness of a rainy night had its difficulties. It was half-past ten before the speedometers registered twenty-two miles. We held a consultation. The *yurts* ought to be a half mile off the trail, if we were right. We turned abruptly eastward and bumped over the rough sagebrush hillocks into the open desert. Suddenly a dog barked; then we saw a flicker of light. We hit the village right in the middle, but what a panic! Men and women ran out, half clothed, screaming in fear. They thought we were bandits. It was ten minutes before they could comprehend who we were.

Then Bato's wife cleared her baby and herself out of the best *yurt,* and brought tea and Mongol cheese. Mac and I slept like dead men while the rain poured outside.

At five o'clock the next afternoon we ascended the escarpment to the vast peneplain that rolled in yellow, waterless waves for a hundred miles to the east. Two miles from camp a lone wolf trotted out of the west to his nightly hunting ground in the basin. He looked huge and gaunt in the slanting sunlight. Puzzled by the cars, he loped slowly away, head turned over his shoulder. I took one shot and he went down, stone dead. That was number thirteen for Wolf Camp. The men were all out to meet us when our cars rolled up to the tents.

"Did you bring the booze?" asked Granger. "We haven't had a drop of alcohol for a week."

First, I had to hear what they had found in my absence; then we told them the news of China, and produced letters and papers. It was July fourteenth, Bastille Day in France, and dinner that night was in honor of our colleague, Père Teilhard de Chardin. The great death trap of shovel-tusked mastodons was exhausted I learned. A dozen jaws and two skulls, many tusks and more skeletal parts had been exposed. They were waiting only for shellac, to take them out.

Already work had been resumed on the deposit just below camp, where only baby mastodons, deer, foxes, and other small mammals had been entombed. The next morning Bill Thomson found both thigh bones of an adult female mastodon. She had died lying on her side, and within the pelvis was the skull and jaw of an unborn baby. Bill performed the operation with Granger in attendance as consulting physician. The rest of us watched and offered gratuitous advice. Also we referred to them insultingly as "palæontological midwives" and "fossil obstetricians" who could only deliver a baby stillborn.

Fœtal specimens of any fossil animal are extremely rare and enormously interesting. Particularly important are they in the Proboscidea, because the teeth of mammoths, elephants and mastodons have a growth and succession quite unlike that of other mammals. The great molars grow upward and forward, pushing out the first teeth, so that only two are in use on either side of each jaw at any one time. Our unborn baby had a jaw about twelve inches long. In the adult the jaw is nearly six feet in length. The two flat incisor tusks projected only an inch beyond the bone, in contrast to the ten-inch plates of a full-grown male.

The collectors also found in the deposit the skull of a deer with remarkable antlers. They resemble a woman's cupped hand cut off at the wrist, and are just about that size. The antler tines are like small spread fingers. Except for the pregnant female mastodon, all the other animals that had been trapped in this ancient bog were small types. Therefore, as I have remarked, we concluded that it was of very soft, sticky mud, but fairly shallow, enabling the larger beasts to extricate themselves.

Exploration to the north and east along the escarpment which marked the former lake shore revealed other bog deposits. One was exactly like the great quarry from which we had removed so many shovel-tusked mastodons. Two or three huge jaws were partly exposed. Doubtless, it would prove to be as rich as the other, if not more so.

But we could not afford to spend more time in that formation, although we had but scratched the surface. Two months were gone already and the season was half over. A request to the Chinese Committee for the Preservation of Cultural Objects, in Peking, for permits to continue the Expedition's researches next year, had met with discouragement. Such permission would be granted only on conditions which no museum could accept.

After five years of arduous exploration of the Gobi Desert, and the expenditure of more than a half million dollars, we had at last discovered the region where we might find primates. The bog and river drift deposits along the shores of this ancient lake may well be the tombs of primitive humans, or near-humans. Primates may have been trapped with other mammals in those far, dim days when the great shovel-tusked mastodons splashed along the edges of the lake. But careful, systematic excavation of every deposit is the only possible way to succeed. Valuable specimens, each one of which reveals a new page in the book of ancient life, will be found whether or not primates are there. That summer we had made a beginning, but only a beginning. Several years of work remain to be done. It was useless for us to undertake a proper systematic exploration, knowing that it could not be completed except under impossible restrictions. Several untouched localities awaited us in the west where we were sure we would find a new fauna and we must at least dip into them before the season ended. Therefore, most reluctantly, it was decided to terminate the work at Wolf Camp.

In the meantime Young and I made another trip to

Peking with two carloads of specimens. The war news indicated that the northern generals probably would get the worst of the struggle and that we had best get our valuable collections to Peking as rapidly as possible. On the second trip we brought back as a camp visitor the late Colonel N. E. Margetts, Military Attaché of the American Legation.

The grassland gazelles had already begun to gather in great herds, and Colonel Margetts saw a spectacle which thrilled him to the toes. It was late in the afternoon and the low sun shot oblique golden paths between the rolling hills. In the shadows of a deep valley we saw a restless mass of yellow forms. Thousands upon thousands of antelope—does, bucks and skipping babies. Colonel Margetts could hardly believe his eyes. It even gave me a thrill, and I have seen many such. Probably nowhere else except in Africa can one find such herds of any game animal. As the cars swung down the trail, the herd moved slowly up the slope of a low hill. Shafts of sunlight caught the leaders, like the spotlights of a theater, changing them to unreal shapes of living gold. They blanketed the hillside and stretched away into the dim mystery of the darkened valley.

There seemed to be an inexhaustible supply of wolves at Wolf Camp. Some one shot a wolf almost every day. One of the Chinese surprised a pair with two young some distance from their den. The mother seized one of the babies, about the size of a cat, in her mouth and made off down a ravine. Colonel Margetts and I shot two half-grown cubs at the same place.

Of course, antelope meat was our chief food, but

sand grouse furnished a welcome variety. They are strange birds, half pigeon, half partridge in appearance. Their three-toed feet are padded like those of a camel, to fit them for walking on the sand. At night and in the morning they make long flights for water.

We used to conceal ourselves near the well, where there was a small pool, and had magnificent shooting. Seventy-four birds in an hour and a half was our best bag. It could be duplicated almost any day.

Lieutenant Wyman was kept busy mapping the little-known region to the east of camp. When he had finished and the baby pit in front of the tents yielded no more specimens, we shifted operations fifty miles due west.

Walter Granger and I stood in front of our tent on the last evening, looking at the great pile of fossil specimens. Then we turned to the shadow-flecked badlands filled with the light of a gorgeous sunset and he said to me, "Roy, we've given the Gobi some of the best years of our lives but the desert has paid its debt."

CHAPTER XX

The End of the Trail

IN 1928, Granger, while on a trip of exploration, had seen a rough escarpment fronting a vast basin. He stopped an hour and found few fossils, but many snakes. The poisonous vipers wriggled out from crevices in the rocks, from behind bushes and from under stones. He came away with a very vivid recollection of the place as being most unhealthy. Still it held possibilities of important fossil discoveries.

So we pitched camp prepared to fight a battle for possession with the snakes. But for some strange reason the reptiles had virtually disappeared. Only five or six disturbed our peace. Granger discovered one curled up under his hat and narrowly missed stepping on another with his bare feet. But that was nothing compared to 1925, when, at Viper Camp, we killed forty-seven in the tents.

Our new home, Camp Margetts, was much like the place we had just left. Behind us was a great plain of gravel and stunted sagebrush; in front, a basin so vast that to the naked eye its northern boundary became only a vague horizon blur of purple. It was like looking over a calm sea from the deck of a giant ship. That night in the silence of the desert twilight we watched gazelles feeding almost at our feet. A hundred or more picked their way daintily among the clumps of sagebrush, nibbling at the tufts of short, stiff grass. Now and then they stood motionless to gaze at the city of cloth which had so suddenly risen on the empty desert. We did not molest these gentle visitors, but shot our meat behind camp on the plain.

It is always interesting to make the first reconnaissance of a new exposure. Usually some specimens are easily found. Each man selects a part of the escarpment and skims the fossil cream. Later, other bones, not so obvious, are discovered. Titanotheres occupied the center of the stage for the first few days. Discovery of a dozen species and genera, all of them grotesque creatures, made us think that we knew the most bizarre animals Nature could produce. Yet almost immediately, at Camp Margetts, a beast with a saddle head was found. It was a new titanothere.

The great *Embolotherium,* the battering-ram titanothere, which we discovered in 1928, also had a saddle head, but this fellow was quite unlike his larger relative.

In some parts of the escarpment there was a top layer of golden yellow sand, evidently a later formation. It was identified definitely by discovering in it the bones of a *Baluchitherium. Entelodon,* a giant pig, was another discovery. Also several new rhinoceroses, one of them with a long, slender muzzle like a horse, and a carnivore, somewhat resembling an enormous wolf, are unlike anything known to science. Probably the latter played havoc among the titanotheres that then roamed the Mongolian plains. Truly this Gobi Desert was a strange place in those ancient days!

Among the most remarkable finds were fourteen soft-shell turtles, related to *Trionyx*. Fossil turtles rather bore me, as a rule. They show so little. Even back in that incredibly remote time of the Age of Reptiles turtles were just turtles. They look almost like the turtles of today. They do not present any very interesting facts of evolution. But these turtles at Camp Margetts were rather special exhibits. Soft-shell turtles—or others, for that matter—that have any parts of the skeleton preserved are extremely rare. These turtles, found by Buckshot, not only had the upper and lower shells but the skulls, vertebræ, limbs and feet almost as beautifully preserved as though the animals had died yesterday. Seven of them lay so close together that they could not be separated and had to be removed in a huge block. The others were taken out one by one. What could have caused so many to die in that single spot? I suppose that they did not all actually die there, but were swept by the current of a stream into some quiet backwater as a final resting place.

One thing that interested me enormously was a quantity of charcoal close to the turtle deposit. They were pieces of twigs about the size of my finger, but many of them. One immediately thinks of human work where fire is discovered. I can hear some one suggest that primitive men had gathered turtles for soup and that the charcoal was the remains of their fire. What a spectacular story it would make if we only had a few facts to support such a thesis. But I am afraid that it can't be done. Man may have existed as far back as the Oligocene as a creature already separate from the ape stem. Professor Osborn thinks so. But it is as certain as anything in the past can be, that no pre-human could have had sufficient intelligence to utilize fire at that early period. The presence of charcoal near the turtles must be ascribed to fire from natural causes, probably lightning. Forest fires today often arise from trees being struck by lightning, and doubtless it happened just as frequently in the past. Quite certainly it was in that way that early humans first learned the use of fire.

One morning, Colonel Margetts, Granger, Young and I drove to the southwest to follow the course of the escarpment upon the edge of which we were encamped. The great bluff extended for many miles, sometimes almost lost in gentle slopes, only to appear again as a precipitous wall, deep bitten by ravines and gullies. Everywhere there were evidences of fossils. Five miles out in the basin we could see a second exposure of red badlands. Colonel Margetts shot a wolf and an antelope before we reached them. Wandering over the brick-red sediments, all of us eventually arrived at an isolated cone-shaped butte. From its low summit we looked down upon a relief map of rounded hillocks, tiny, flat plains and miniature ravines. Almost every inch was covered with animal footprints. An intricate tracery of innumerable lines crossed and recrossed the smooth red surface. Just below us a herd of gazelle had wandered out from the mouth of a greater ravine. Their delicate, pointed tracks showed clear and distinct. An antelope had been chased by a wolf around the base of yonder hill. A skull and bleached bones showed where a gazelle had been devoured. Foxes, following hard upon the six-foot leaps of kangaroo rats, had left their story. The track of a hare, going slowly, led to the edge of a shallow ravine, but the little creature had left in sudden panic. I wondered why. Close under the bank, in soft sand, was a wolf bed. No wonder that the hare had run for its life. Feathers of a sand grouse were scattered at one spot, but no tracks showed. The attack had come from above. A falcon, probably, had struck in the air, dropping like a bullet from the clouds.

For an hour I wandered over the badlands reading the story of life and death in the desert. Sixty million years ago, when these red sediments were being deposited, the drama was the same, but with different actors. Then rhinoceroses trod this ground; the gigantic *Andrewsarcus,* greatest of all known flesh eaters, prowled at night, and hyenas fed upon the bodies of dying titanotheres. It was a world of nightmare creatures, but the vegetation was not so strange. Then the desert did not exist. The high plateaus of Africa today, with their open plains and sparse forests, offer a convincing parallel to ancient Mongolia.

At one spot in the sediments I found bone fragments thickly scattered on the surface. Granger marked it for future investigation and it proved to be a veritable gold mine. Some unexplained natural cataclysm had overwhelmed a whole herd of Chalicotheres. They had all died in this spot or been carried there by water. Skulls and skeletons were packed in a solid bone layer. The Chalicothere was a "clawed-hoofed" mammal, a veritable paradox. The head resembled that of a horse, but the hoofs were replaced by enormous claws. Palæontologists have puzzled for years about the meaning of those claws and hardly a reasonable suggestion has been forthcoming.

Other Chalicotheres are known in America and Europe, but we had found only jaws in previous years.

On August twenty-eighth Mac Young and I left for Kalgan on the fourth trip, with two carloads of fossils. I was to stay in Peking to carry on diplomatic negotiations for the continuation of our explorations. Granger took charge of the field work of the Expedition and expected to remain in Mongolia for another month. The war had progressed so unfavorably for the northern generals that Yen Hsi-shan needed every man at the front. He withdrew the soldiers who had been patrolling the Kalgan-Urga trail, and immediately bandits returned to their old hunting grounds. In Chapter VII, on the life of McKenzie Young, I have told of our trip down and his exciting return when he was attacked by bandits.

Mac and Liu reached camp at nine o'clock on the evening of the attack. Granger had expected to move to a new place, but the region they were in was too rich to leave. More Amblypods were discovered and they obtained some of the most valuable specimens of the entire expedition. It was not until the end of September that they broke camp.

The camels were sent to Hatt-in Sumu with the specimens. The men went by motor. Winter came with a rush, as it usually does in Mongolia. A blizzard raged all the afternoon and night of September thirtieth, burying the country in snow. It was high time to be gone.

Lucky it was that the Expedition moved out when it did, for things began to happen on the road the day after their departure. Bandits had captured five Chinese merchants and held them for fifty thousand dollars' ransom.

In retaliation, the "soldiers" took the father and brother of the chief brigand. This started a real warfare, for isolated groups of bandits consolidated into a force of four or five hundred. The road police hastily retired from Chap Ser, leaving the robbers in possession. Our cars were the last to pass on the trail.

Mac Young remained in Changpei-hsien, thirty-four miles from Kalgan, to receive the camels which were bringing our fossils. After a long delay, they arrived, and by means of the customary outlay of silver in "squeeze," he got them to Peking. The fossils, spread out upon the laboratory floor, were an impressive sight. It was the largest and perhaps the most important collection ever taken out of Central Asia. The dinosaur eggs and some other specimens discovered in previous years were more spectacular and had more popular interest, but from the standpoint of pure science they hardly surpassed this collection.

It gave us additional proof that the Central Asian plateau was one of the greatest centers of origin and distribution of animal life during the Age of Reptiles and the Age of Mammals. It shows that Mongolia was even more favorable for the development of many types of mammals than was Europe or America and continued to be so long after these groups had disappeared in other parts of the world. It gives much additional knowledge of the climate, vegetation and physical conditions of this great incubator of world life.

True, we have not been successful in one objective of our search—the dawn man. It is a scientific tragedy that Chinese opposition to foreign investigation should end our work just when that goal might be attained. Still we have shown the way; have broken the trail, as it were. Later, others will reap a rich harvest. We are more than ever convinced that Central Asia was the palæontological Garden of Eden. Future work will demonstrate whether we are right or wrong.

APPENDIX A

The Dune Dwellers of Mongolia

DURING our explorations southeast of Mastodon Camp at the end of the 1928 Expedition we pitched the tents one evening just above a "blow-out" or wind hollow in the bottom of which was a well of excellent water. Thomson found that the hollow had been a great Dune Dweller site. Hundreds of artifacts were scattered about and best of all bits of bone were discovered *in situ.* This was the first Dune Dweller station where bones were found.

We spent that evening and all next day prospecting the Dune Dweller station which lay so conveniently close to our camp. Right in the middle of the path which led down into the bottom of the well, an ancient hearth was discovered. The earth, fire-blackened and hardened, was unmistakable. On the earth itself we found frog and bird bones and in the immediate vicinity what evidently had been a necklace of fox canine teeth neatly drilled through the bones. Also small fresh water clam shells were drilled and had been used in a similar way. Bones of wild ass, hares, gazelle and many birds were discovered. Some of the larger bird bones were decorated with parallel lines beautifully etched. But not a trace of human bones; *metates,* pestles, scrapers, hammer stones and arrow and spear points were numerous. There was also much broken pottery which evidently had been made by molding the wet clay in a basket. In the vicinity other accumulations of artifacts were found showing that this must have been the site of a considerable village. It was quite the most important station that we have discovered.

It is most surprising that no human remains have been found near any of the Dune Dweller stations. Evidently the conditions were not proper for the preservation of bone or we should have discovered many animal remains. These people subsisted largely upon game, and of course, hundreds of thousands of animals and birds must have been eaten. Yet except for this one spot no bones have been discovered. It is entirely probable that their dead were buried at some distance from their village sites.

Although the Dune Dweller culture is somewhat similar to the Azilian of Europe, it differs from it in many important ways. It appears to be distinctly Gobian and not closely related to any known cultures in other parts of the world. Where did they come from and where did they go? Only extended explorations of contiguous regions can give an answer to the problem. What the Dune Dwellers looked like we do not know. That they were hunters there can be no doubt. Certainly they dressed in skins, for Mongolia was cold in winter even then. But it was very much less arid and the Gobi was by no means such an inhospitable desert as it is today.

The Dune Dwellers could not have lived in caves for caves are virtually non-existent in Mongolia. They must have built shelters out of skins on the sunny sides of banks or dunes. The roots of the tamarisk and other stiff vegetation offer obstructions to the wind-blown sand and about them dunes are formed. Almost every Gobi lake has dunes on at least one side. Thus our people chose such spots as permanent camps for they had the essentials of life—fuel, water and comparative shelter. In the very earliest stage of their culture we found no arrow or spear points. Presumably they made them of bone which has not been preserved. In later deposits small beautifully worked arrow points were plentiful.

We have already found many great residence sites of these people in Mongolia. Of course, there must be hundreds yet undiscovered. At first we thought that they were few in number and had a limited distribution. Now we know that the opposite is true. *Twenty thousand years ago Mongolia was much more densely populated by these Stone Age people than it is today or probably has been during historic times.* Even when Genghis Khan conquered all of Asia and much of Europe, it is doubtful if he had half as great a population from which to recruit his armies as existed in Mongolia where the Dune Dwellers lived.

We have explored most of central and southern Mongolia and everywhere this culture appears. Probably the Dune Dwellers were grouped in communities where living conditions were most favorable and made excursions of short duration into the more arid parts of the desert. Certainly they were a hardy people of considerable strength and endurance. No weaklings could have lived in such open country under the semi-desert conditions, and severe climate. Hunting was difficult then as it is today. Great skill in stalking was required even though game was abundant. Gazelle and wild ass probably formed their principal food, reënforced with smaller animals such as wolves, foxes, hares, badgers, kangaroo rats and other rodents. In the mountains they could get ibex and wild sheep. As some of the country was undoubtedly sparsely forested wapiti, roedeer, bear and perhaps moose were to be had. Birds they could catch in nets and snares, but to obtain such food they must exercise a good deal of intelligence and skill. That in the later stages, the true Neolithic, they had a certain vegetable diet of sorts, we were surprised to learn by finding grinding stones and rollers. They may have used these to break up roots and seeds but it is possible that certain kinds of wild grain then grew in Mongolia under the less desert conditions.

We have discovered no traces of any art. That is hardly to be expected from a plains living people. Art was only developed during the period of cave life when there was opportunity for contemplation. Yet, the skill which the Dune Dwellers exhibited in making their artifacts is quite remarkable. Except in a few places the materials were not good. They utilized whatever stone was at hand, and we find implements made of quartzite, churt chalcedony, jasper and half a dozen volcanic rocks. They could produce the most exquisite flakes which were used as knives and drills. It is not certain whether they obtained these by steady pushing against a core or by a sharp blow with a hammer stone. As a whole their culture is microlithic; that is, the scrapers, knives, drills and arrow points are very small and delicate.

I believe it is improbable that the Dune Dwellers were directly ancestral to the tribesmen who inhabited

Mongolia before the present Mongols appeared. It is more likely that they left the country as the increasing aridity converted more and more of Mongolia into an inhospitable desert. Lakes and rivers began to dry up and the scanty forests to disappear, game was less abundant; in short, the plateau became an impossible residence for any primitive people. They were forced to migrate and migrate they did, carrying their culture to new lands. Thus it might easily have reached Europe. I have remarked that geological evidence appears to place the Dune Dwellers as considerably older than the Azilians, their nearest cultural representatives in France and Scandinavia.

During the summer of 1934 Dune Dweller artifacts were discovered in Alaska. It is evident, therefore, that these people migrated to a considerable distance. What their relationship, if any, is to the North American Indians remains to be demonstrated.

APPENDIX B

Summary of Outstanding Results of the Central Asiatic Expeditions

BEFORE the Central Asiatic Expeditions went to the Gobi Desert, Mongolia was virtually unknown scientifically. We were told that it was a desolate waste of sand and gravel signifying nothing. Desolate it is in all truth, but we found it to be a veritable treasure house packed with unknown riches. Those years of work "revealed a new volume in the history of the earth," to quote Professor Henry Fairfield Osborn, whose brilliant prophecy it was that sent us there.

We have far from exhausted its possibilities; in fact we have but scratched the surface. Yet the information that it has already yielded supplies many unknown links in earth history and gives a clew to what still remains to be discovered. We have learned that *Central Asia is the oldest continuously dry land in the world.* In other words, when Europe and America were still being periodically elevated and submerged, Central Asia was dry land and has remained so since the middle of the Age of Reptiles.

Underneath the plateau there is an enormous granite bathylith, probably the largest in the world, which acts as a floor on which the later sediments have been deposited. We have found that the plateau *never was covered by ice* during the period when glaciers were periodically descending and withdrawing over the continents of Europe and America. In Central Asia there were only small glaciers in the mountains which did not extend far out on the plains themselves. In Pleistocene time glacial epochs were represented in the Gobi region by a change to more humid climate, and interglacial epochs by return to desert conditions.

The Gobi Desert is essentially a rock desert with a very thin veneer of shifting sand and much bare rock. Evidence is conclusive that this region of *Central Mongolia has been arid and semi-arid in its climatic habit for many millions of years, but there have been cycles of greater and less aridity.* The climate, even in recent times, had not been uniform. There have been smaller changes within the longer cycles. Even within the last great arid cycle there have been epochs when the Gobi region was comparatively fruitful and could support a numerous population, separated by different epochs and difficult conditions for animal life.

We know now that *man occupied the valleys of the Gobi region immediately after the close of the glacial period,* and that the different cultures of prehistoric man which we discovered probably correspond to recurrent favorable climatic conditions. In the intervening times conditions were too severe for these peoples to survive.

We have discovered that *this plateau was the mother of the continental life of Europe and America.* It was a sort of palæontological incubator where great groups of reptiles and mammals got their start in life and spread to other parts of the world. That Mongolia is one of the world's greatest fossil fields is evident.

Appendix B

It was like a dream come true—the way we discovered animals science had told us must be there; also many others that no one had even thought of. Just glance at some of the thousands of fossils in our collections. There are the *dinosaur eggs*. There are more than *a hundred skulls of skeletons of dinosaurs,* previously unknown to science, most of them representing ancestral types. In some the preservation is wonderful; probably the finest of any known specimens. These reptiles lived eighty or ninety million years ago, and yet in one or two we could even see the outlines of the stomach and impressions of the skin in the matrix of red sandstone. Then, we have *the only skulls in the world of the oldest true placental mammals.* They were tiny creatures not much larger than a rat, yet they represent Nature's first attempt to populate the earth with higher types of animals than the cold-blooded dinosaurs which had dominated it during the Age of Reptiles. Those seven little skulls did not mean much to the man in the street, but they are probably the most important single specimens in our entire collection.

Also we found the *skull and parts of the skeleton of Baluchitherium,* the world's largest mammal. There is, too, *Andrewsarchus, greatest of all flesh-eating mammals.* These are only a few of the spectacular exhibits, but still more important are the teeth and jaws and skeletons which *represent types ancestral to animals of Europe and America, India and Africa.*

As Professor Osborn has written in the Encyclopædia Britannica, "these discoveries have established Mongolia as a treasure-house of the life-history of the Earth from the close of the Jurassic time onward to the close of the Pleistocene time, *revealing especially the hitherto unknown high continental life of Cretaceous and Tertiary time.* Consequently, the outstanding geologic discovery of the Expedition is, first, that Gobia since Jurassic time has been a central Asiatic continent extremely favorable to the evolution of reptiles, mammals, insects and plants hitherto known only along the Cretaceous shore-lines of Europe and the Cretaceous sea-borders of the center of America; and, second, that this now terribly desert region, traversed by the gazelle and the wild ass, was certainly luxuriant with life throughout Cretaceous and Tertiary time, sparsely forested, with limited rain supply like the high plateau regions of Africa today."

In zoölogy, the Expedition *collected approximately ten thousand mammals, eight thousand reptiles and amphibians and eight thousand fish* from China and Mongolia. In addition to the description of many types new to science, the ranges and relationships of numerous species have been delineated and clarified. The collections have indicated the chief faunal divisions in Mongolia and China. For the first time, sufficient material from these regions has been assembled to give a general view of the faunas and to make possible handbooks of mammals, reptiles and amphibians and fish which are now being written.

The *five hundred species comprising the botanical collection* give a comprehensive view of the flora of Mongolia south of the northern forests. Further studies of the distribution of the living plants will, undoubtedly, throw light on the past climate.

Appendix B

A comparison between the climatic conditions of Mongolia and that portion of China lying immediately to the south during the Age of Reptiles and Age of Mammals brings out the fact that *the relative difference in rainfall which now exists between these two areas has probably extended far back into geologic times.* On the basis of the fossil record, it is evident that *a Sequoia forest, closely related to the giant redwoods of western America, occupied the windward side of the Khingan Mountains and other ranges which run across northern China.*

The living trees of northern China and Japan are of special interest because of the similarity of the leaves of certain species to those of fossil species in the Tertiary of western North America. A species of hawthorn, *Crataegus pinnatifida,* has leaves which closely resemble those of the fossil *C. newberri* of eastern Oregon, and there is no living American species which shows a like relationship. One of the common elms of northern China, *Ulmus parvifolia,* is much like a fossil elm from eastern Oregon which has no near relative, either living or fossil, in America.

Interestingly enough, the Sequoia, which is now limited to western North America, is found fossil in Manchuria. We therefore have the situation of certain kinds of trees which occur in North America as fossils but no longer live there, although they are present today in the forests of Asia; and there are others, which, though now extinct, are shown by the fossil record to have lived in eastern Asia in the Tertiary, and which survive in the living forests of America.

During the Expeditions' work, *about fifty thousand*

feet of motion-picture film and many thousand still photographs were taken. These record all aspects of the Expeditions' activities, as well as almost every phase of Mongolian life. Such a permanent record of a rapidly vanishing culture is of great value.

While we were in Mongolia, Major Roberts developed *an entirely new method of mapping, adapted to a plains country where there are few natural landmarks.* By this method a continuous traverse of nearly a thousand miles was made northwestward through the heart of Mongolia. The elevations by which the topography was sketched were based on an instrumental vertical angle line. This was the first time that this method was used on the Central Asian plateau. Secondary traverses were run by prismatic compass for many hundreds of miles and detail maps made of special areas. Altogether forty-four maps, almost all of them representing new areas, were made and a general map of Mongolia, which is just ready for publication, has been prepared.

Just as a mass of theoretical evidence indicated that Central Asia was a great origin and distribution center for lower animals, so *it is also indicated as the birthplace of the human species.* Inevitably, the human history has come to occupy our thoughts more and more. One story we want is that of the Old Stone Age cultures in Central Asia and their relation to those of Europe and Africa. But our greatest interest is in the incredibly more ancient problem of the origin and development of man; a story which extends into the past for perhaps millions of years.

Most of the ranking authorities today agree that man must have originated in Asia, but the part of Asia is in dispute. We believe that our high plateau was the so-called Garden of Eden. Not that it looks much like a garden today. But, as I have said, it was not always thus. A few hundreds of thousands of years ago, Central Asia was probably not unlike the high plateau region of Africa at the present time. Semi-arid in spots, there were also rolling meadowlands and patches of open forest. The climate was cool and exhilarating; game abundant. Conditions of life were just difficult enough to stimulate effort both mental and physical. Such is the ideal environment for the development of the early human type as visualized by Professor Osborn.

Almost certainly man could not have progressed far up the evolutionary ladder in a tropical or heavily forested country. Life there was too easy. Fruit and nuts were abundant; he had but to stretch out his hand to obtain food enough to last him for days. The effort of living was reduced to a minimum. Early man, as a type, required the stimulus of mental and physical effort for development, just as man individually requires it today. The tropical forests a million or two years ago were regions of retirement for three types defeated in the struggle for existence by more virile competitors; they could not have been incubators for the dominant forms of mammalian life.

We know now that the human species is incredibly more ancient than it was supposed to be thirty years ago. Then, four or five hundred thousand years was set as the uttermost limit; now it is estimated at several million years. With each new discovery man's age is being pushed back further and further into the past. One reason is the more exact methods of determining the ages of geological strata; another is the discovery of flint implements almost certainly worked by human hands in the Pliocene period of the Age of Mammals.

Hunting for the bones of primitive man in any part of the world is the greatest gamble in scientific exploration. Human remains are much more difficult to discover than are those of other mammals. In the first place, their numbers were infinitely few compared to the teeming millions of lower creatures. In the second place, even at such an early stage, the dawn men were more intelligent than their contemporary animals and more frequently avoided the quicksand traps, marshes and streams which form the most prolific fossil deposits. In the third place, their bones were so delicate and easily broken that they were not so often preserved as were the more compact skeletons of other animals. Still, some have been found; others must eventually come to light if the search is continued. Our method is to find and study with the utmost care those geological strata which are of such an age that they could contain human types. We bring to bear upon the problem the best scientific knowledge and experience available; the result is in the lap of the gods.

CAPTAIN HILL SURVEYING AT BALUCH CAMP

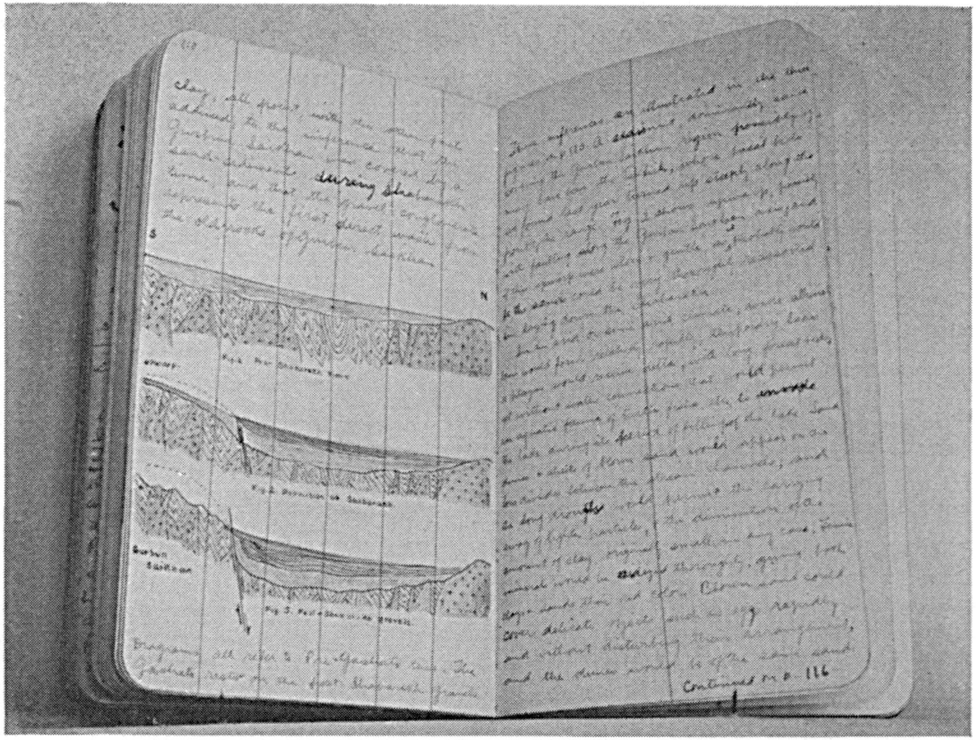

TYPICAL PAGE FROM THE GEOLOGISTS' DAILY ROUTE NOTEBOOK

Professors Berkey and Morris carried their geological sections in this way throughout the entire route of the Expedition

BANDITS HALT THE EXPEDITION'S CARS

READY FOR AN EXPECTED ATTACK BY BANDITS

ONE OF THE HAZARDS OF GOBI TRAVEL.

OUR PET ANTELOPE NURSING FROM ITS FOSTER MOTHER, A MONGOL GOAT

HEAD OF DESERT GAZELLE (*Gazella subgutturosa hillieriana*)

HEAD OF MONGOLIAN BLACK VULTURE

PUSHING ONE OF THE EXPEDITION CARS THROUGH THE SAND. WOLF ON TOP OF THE LOAD

A WELL IN THE DESERT

MONGOLS LISTENING TO VICTROLA

A PRAYER WHEEL AND A LAMA AT URGA

A FIVE-YEAR-OLD MONGOL SHEEP HERDER

MCKENZIE YOUNG AND PET ANTELOPE

MONGOL DRAWS MAP IN THE SAND OF WESTERN TRAILS

EXPEDITION IN CAMP AT NIGHT. McKENZIE YOUNG THIRD FROM RIGHT

ANDREWS AND TSERIN GO OUT TO MEET THE CAMELS

ANDREWS, GRANGER AND YOUNG REPACKING CARAVAN SUPPLIES

DR. ANDREWS SCANS THE DESERT FOR TRAILS THAT LEAD TO THE EAST

MONGOL YURT AND CORRAL FOR LAMBS, KIDS AND CALVES

TSERIN POINTS OUT APPROACHING CARAVAN TO ANDREWS

PATCHING A CAMEL'S FOOT WITH RUBBER TIRE

FOSSIL-BEARING COUNTRY AT BALUCH CAMP

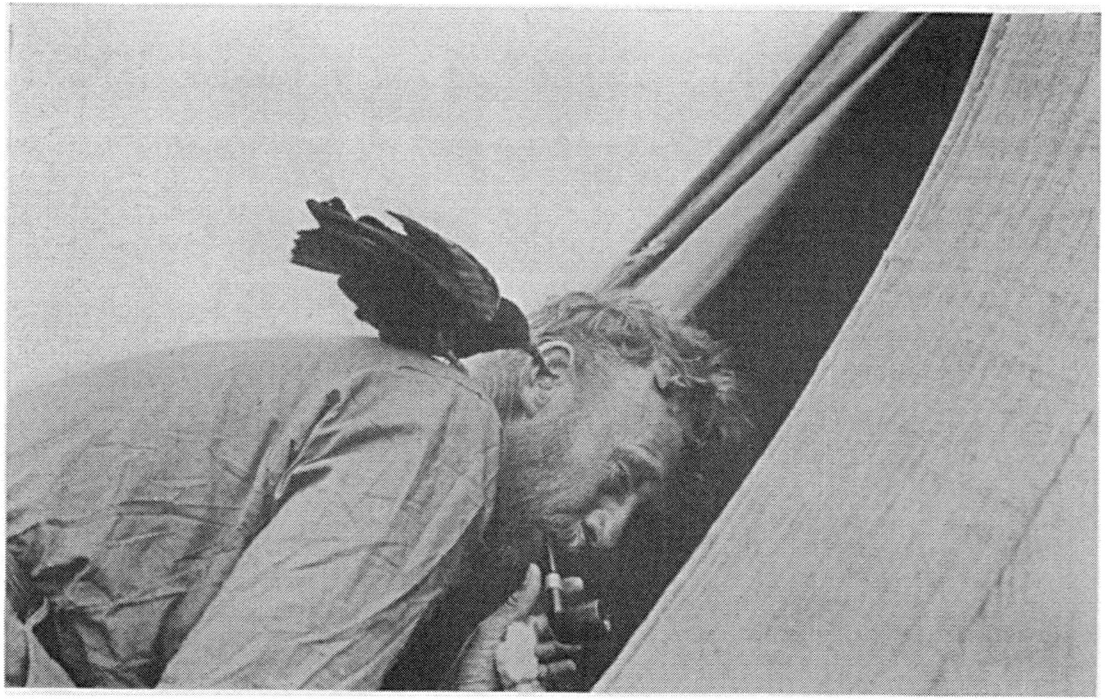

A PET RED-BILLED CHOUGH TELLS THE CHIEF PALÆONTOLOGIST A SECRET

DINOSAURS AT IREN DABASU (ERHLIEN)

THE SITE OF AN ANCIENT DUNE DWELLER HEARTH

Remains of fire and of bones were found on the hearth

THE EXPEDITION AT "WOLF CAMP"

Entelodon, A GIANT PIG-LIKE ANIMAL
Painted by Charles R. Knight

RESTORATION OF *Baluchitherium*
Painted by Charles R. Knight

GRANGER EXCAVATING THE SKULL OF AN *Embolotherium*

RESTORATION OF SHOVEL-TUSKED MASTODON BY MARGRET FLINSCH

TWO JAWS OF SHOVEL-TUSKED MASTODONS FROM THE GREAT QUARRY

At the extreme right of the picture may be seen the broad flattened tusks of another pair of jaws

GRANGER AND THOMSON WITH A SHOVEL-TUSKED MASTODON'S JAW

This specimen is still in its field casing of burlap and paste

THE GRAVEYARD OF THE SHOVEL-TUSKED MASTODONS AFTER THREE WEEKS' EXCAVATION

Sportsman's Vintage Press

Available at www.SportsmansVintagePress.com

Sixguns..Elmer Keith
Sixgun Cartridges and Loads..................................Elmer Keith
Big Game Rifles and Cartridges..............................Elmer Keith
Shotguns by Keith..Elmer Keith
Elmer Keith's Big Game Hunting............................Elmer Keith
Keith's Rifles for Large Game.................................Elmer Keith
The Secrets of Double Action Shooting...................Bob Nichols
Wilderness Hunting and Wildcraft...........................Townsend Whelen
Why Not Load Your Own..Townsend Whelen
.22 Caliber Rifle Shooting.......................................C. S. Landis
Rifle-Craft...C. S. Landis
The American Deer Hunter......................................Francis Sell
Asian Jungle, African Bush.....................................Charles Askins
Modern Shotguns and Loads...................................Charles Askins
The African Hunt...Charles Askins
African Rifles and Cartridges...........................John "Pondoro" Taylor
Smith & Wesson Hand Guns...................................McHenry & Roper
Hunting Pronghorn Antelope..................................Bert Popowski
The Woodchuck Hunter...Paul Estey
Small Bore Rifle Shooting.......................................Edward Crossman
Modern Bow Hunting..Hiram Grogan
Salt Water Fishing Is Easy.......................................Jerry Sylvester
Fishing Flies and Fly Tying.....................................William Blades
This Business of Exploring.............................Roy Chapman Andrews
Handbook for Shooters and Reloaders......................P. O. Ackley
Greatest Fishing...Joe Brooks
Hunting Our Biggest Game.......................................Clyde Ormond
How To Hunt Deer...Edward Freeman

All books are available to order as hard copy and some books are available as a free PDF download. Visit **www.SportsmansVintagePress.com** for a full description and chapter listing of each title and follow us on Twitter **@SVPress** for updates on sales and new releases.

Made in the USA
Middletown, DE
09 June 2021

41680256R00205